The Omai Files

RONALD SELVON SEALES

authorHOUSE®

AuthorHouse™
1663 Liberty Drive, Suite 200
Bloomington, IN 47403
www.authorhouse.com
Phone: 1-800-839-8640

First published by AuthorHouse 3/3/2008

ISBN: 978-1-4343-5694-9 (sc)

Printed in the United States of America
Bloomington, Indiana

This book is printed on acid-free paper.

To my parents Kenneth and Lillian Seales of Lake City, S.C.

There is a tide in the affairs of men, which taken at the flood leads on to fortune;

Omitted, all the voyage of their life is bound in shallows and in misery.

On such a full sea are we now afloat:

We must take the current when it serves or lose our ventures.

William Shakespeare...
from Julius Caesar

Author's Note

While the cyanide spill in the Omai River actually occurred in August of 1995, *The Omai Files* is nonetheless a product of my imagination. Similarly, while the names of Tribes of Guyana's indigenous peoples are very real, characters portrayed in this novel are fictitious. Also fictitious are all other characters. The author's mention of agencies, organizations or geographical regions is in no way intended to represent reality or actual specific issues.

RSS

Chapter 1

Guyana 1995.

It was late October. About fifteen miles west of Ituni, far in the Essequibo rain forest, there was a town. It lay sandwiched between the Essequibo and Omai Rivers, south east of Potaro Landing.

It bore the tragically unfortunate name of Omai.

Omai was dead.

The mortal blow was inflicted by a major Canadian mining company. The Wadpole Corporation. Guyana's largest economic contributor after the demise of the sugar cane industry.

It happened one night in August, 1995. And days later, the news hit Georgetown, the capital, like a monstrous volcanic eruption.

Four billion liters of cyanide-laced slush was released into the giant Essequibo River and its tributaries, when a dam broke at the company's plant in the town of Omai.

The region of the Essequibo was locked in a chemical holocaust. Such a disaster was never before seen since the infamous Jonestown Massacre of 1978, where more than a thousand people perished.

The impact of the cyanide spill that day was beyond comprehension. Death and destruction were unprecedented. People complained of multiple somatic ailments, including nausea, vomiting and diarrhea before toppling over to their deaths.

Communities downstream as far as Kurupukari, Maipuri Landing and Apoteri reported seeing a reddish slick in the Essequibo, Siparuni and Burro-Burro Rivers. Besides fishkill in the tens of thousands, dead bloated bodies of capabaras, tapirs, manatees and other aquatic life were seen floating down the rivers.

The water was unfit for use.

Residents of the surrounding area were concerned about the short- and long-term effects that cyanide and other chemical pollutants would have on their daily lives. The Wadpole Corporation vehemently denied the dumping of thousands of gallons of chemicals in the Omai River.

Two months later, residents of indigenous Indians communities along the Essequibo region traveled to Georgetown to urge the Guyana High Court of Appeals to reinstate a class action suit filed by the Amerindians in August, 1995 against the Wadpole Corporation. That suit alleged that the owners of the gold-mining operation allowed a dam to collapse pouring several billion liters of cyanide-laced effluent into the Omai River. Thousands of Amerindians assembled on Robb and Alexander Streets, headquarters of Wadpole, as the Guyana High Court refused to hear the case, on so-called technical grounds.

But leaders of the demonstrators urged the government of President Charlie Jagnarine to deal expeditiously with the matter.

One leader told the throng during that historic gathering that Amerindians were there to lobby for what was their right and to show solidarity for their disenfranchised brothers and sisters all across Guyana. He told the gathering that they, the Amerindians, were not seeking political office and those Guyanese who wanted to denigrate them could go right ahead. He said their initiative was an act to demonstrate that Amerindians had rights like all other citizens of Guyana and that they were expressing their dissatisfaction of the way the government had handled the Omai incident.

Speakers after speakers were heard. Then Burt Samuels, a University of Guyana graduate and chief of the Guyanese Organization of Indigenous Peoples, told the crowd that discriminatory practices against Amerindians were seeds that have been sown decades ago. He reminded his audience that an Amerindian Commission for the Preservation of the Environment was set up thirty-six years ago by the then ruling Peoples National Congress, but all its entreaties fell on deaf ears. Preservation of Indian land had never been a priority.

"President Jagnarine has taken our land, the land of our ancestors, and arbitrarily given out concessions to foreign miners and logging corporations without talking to us. The consequence is Omai. Our people are still dying after the cyanide spill two months ago, and no one is taking responsibility. We were not even compensated. Today the sub-human treatment of the Amerindian people by the government of the Peoples Progressive Party continues."

Samuels said that the International Convention of the United Nations and the Organization of American States on indigenous peoples mandated the right to land was based on the occupation and usage of such land to procure subsistence.

"The Wadpole Corporation does not belong in Omai," Samuels said. "It should be made to clean up the area and compensate residents for the harm that was done," he continued. "Both the Wadpole Corporation and President Jagnarine have made promises that Omai's business operations were safe but we all know that people are still dying as far northwest as the Puruni River and as far south as the town of Kuyuwini. I call on my fellow Amerindians and all Guyanese, including the government and the international community, to help us fight for rights that are natural, human and environmental."

In answer to Samuels' exhortation, the crowd cheered and chanted, "We want action!"

Goaded by the people's response, Samuels continued.

"Based on studies by the Pan American Health Organization, our rivers are now polluted. In the interior of our country, the river is everything to our community. It's a system of transport bringing people to each other. No electricity, plumbing or telephones exist. Farming is done for subsistence, not for commerce, but today, plantain, banana, yam and other edibles cannot be sowed owing to polluted soil. Who can compete against the Wadpole Corporation with cutlasses and hoes? No one!"

After a pause, Samuels held his audience with a stubborn gaze and continued.

"Let me say this. We have a problem on our hands. A small amount of two percent solution of cyanide can kill a human being. Those Wadpole Corporation executives and PPP ministers obviously are unaware that I did my homework on cyanide. The chemical blocks the absorption of oxygen by cells, causing the victim to suffocate. That's what Mr. Jagnarine is doing to us all. Suffocating us."

"Do something now! Do something now!" the crowd chanted.

Samuels gestured with appreciation.

"After the spill, we Amerindians met with officials of the Guyana Government and the Wadpole Corporation and demanded compensation and held the mining company responsible for environmental cleanup liabilities. We also demanded zero discharge of pollutants in the Essequibo River or its tributaries as mandated by the United States Environmental Protection Agency. But our cries were ignored by the company in Omai. Even the High Court of Appeals has refused to hear our case but we now call on the United States EPA to conduct a thorough investigation into the Omai disaster and, after bringing the perpetrators to justice, we want to see the mine shut down for good."

The following week, Burt Samuels' group, Guyanese Organization of Indigenous Peoples (GOIP) launched a national signature campaign to force President Jagnarine to reopen the class action suit brought against Wadpole Corporation.

One opposition party parliamentarian Peter Issacs at a meeting in support of the GOIP told his listeners that he would not allow his party to let up its quest to have the mining company investigated.

Issacs said that although the Wadpole Corporation was the largest contributor to Guyana's economy after the sugar industry, yet the present disaster has raised the level of fear not only in people around the Omai area and downriver but throughout the entire country.

As the months went by, President Jagnarine refused to hold a dialogue with members of the GOIP.

The signature campaign intensified.

The GOIP was seeking to compile thousands of signatures for a petition document that was to be forwarded to the United States EPA.

That was a consequence of the daily picketing activities outside Wadpole's headquarters at Robb and Alexander Streets in Georgetown. Not only Amerindians manned the demonstration lines. There were Guyanese of all race, color, creed, economic background and national origin.

In a full page advertisement carried under the Heading, "Governmental Idiocrasy," in the Independent Lacytown News, the GOIP said the Omai incident as a whole "infringes upon the core of our constitutional protections, human rights, system of culpability, political accountability and our personal and national welfare." The statement continued, "It threatens the very survival of future generations."

The GOIP leader said airing his concerns with President Jagnarine directly was virtually impossible. He hid in his burrow on Vlissengen Road like a labba. Just to avoid the reality of Omai.

In addition Samuels said the Omai issue had come to light in international communities. Guyanese Organizations overseas have written foreign embassies in Georgetown to reveal the Omai incident to their respective governments.

Samuels said the spill's impact included economic, ecological, environmental, social and political consequences. Barbados, Trinidad, Jamaica and Antigua had imposed a ban on importation of Guyana's fish. Thus a regional problem was evident.

Chapter 2

Desmond Bobo's final days were spent on a first floor ward at Mahdia Regional Hospital, where the view of rusting zinc rooftops gave way to a polluted Omai River. And breathtaking orange sunsets.

When he was brought by truck from Tumatumari, Desmond Bobo, spiritual leader of the Waikas, one of Guyana's main groups of indigenous Indians, was assigned Ward 105——considered the ward of death.

It was an alternative medical ward where people brought their loved ones seeking hope, with conditions that most doctors deemed incurable. At fifty-eight and partially paralyzed from drinking cyanide-laced water from the Omai River, Bobo was clinging to life. He also had complications from leptospirosis, a bacterial infection caused by ingesting water tainted with the bodies of decaying animals.

"He was really going down," the physician at Mahdia Regional Hospital would later tell officials from the Ministry of Health. "Vomiting blood, with a body temperature of 104°. He was going down fast."

Bobo's ward faced away from the banana fields that once defined that stretch of the Omai coastline, five miles east of Tumatumari. From his window, he would not have been able to see the donkey carts pulling solitary farmers towards their daily tasks, or the stray dogs that roamed the dusty streets, dotted with half-built wooden houses. The only hint of infrastructure used to be the constant screaming of motor boats moving up and down the Omai River.

The hospital, a white painted, two-story facility with long rectangular windows, was built by the British during colonial times. It attracted the aged, the chronically ill, emergencies and snake bit victims.

It was on Ward 105 where Desmond Bobo slipped away as he was being infused with 0.9 normal saline.

As the days unfolded, Mahdia Regional Hospital was inundated with the sick, the dying and the dead due to the Omai cyanide spill.

One news report claimed that the Omai incident was responsible for over four hundred deaths due to cyanide poisoning and leptospirosis.

Chapter 3

Charles Jagnarine, president of Guyana, was an aging man, with a small commodity of grizzled hair. Burnt almond eyes, bulging from their sockets, were hacked into an ancient cadaverous face. The Omai incident had hollowed out his cheekbones and mottled his skin. His sun-baked lips were knotted into a tight permanent line.

Jagnarine was considered by most an old political battle axe. He was born on the Courentyne coast in the county of Berbice. Having been descended from a servile labor force of Indo-Guyanese, he worked his way from farmer boy to dentist. In 1952, he formed the People's Progressive Party with the aid of a young Afro-Guyanese attorney named Lancelot Brown. Together they agreed that Guyana was exploited under its system of colonialism by Great Britain. They set out to dissolve the present system.

The following year the PPP, under the leadership of Charles Jagnarine and Lancelot Brown as second in command, defeated all other political parties in a general election. Jagnarine and Brown formed a government with 80 percent of the electorate. But Jagnarine had fallen asleep and grew too complacent with his own political agenda. He underestimated the political ambition of Lancelot Brown. Then to Jagnarine's astonishment, Brown quickly formed the People's National Congress or PNC. Brown, with a hard, pinched facial expression declared that the PNC was a party created to enhance the political empowerment of Afro-Guyanese.

Jagnarine quickly summoned his followers to Freedom House in Eve Leary. With the breath coming raw in his throat and his face growing haggard with worry, he declared that Lancelot Brown had abandoned the PPP to form his own political party, the People's National Congress. But Jagnarine vowed to fight Brown in the political arena until one gladiator was standing.

Years later, in 1958, both major political parties clashed in a general election. But Charles Jagnarine won the election retaining his title as prime minister of Guyana.

But early in 1960 news reached London and Washington that Charles Jagnarine was closely aligned with Cuba, the USSR and North Korea. And that a socialist republic in Guyana was in the making. That piece of information led to multiple behind the scene activities in London and Washington. Great Britain quickly suspended the Guyana Constitution and a temporary administration was appointed.

A revised constitution was introduced in 1964. And in elections held under a system of proportional representation at years' end, Charles Jagnarine won the largest number of seats in the Legislative Assembly. But he failed to gain a majority to form a government. A coalition government was formed by the PNC and a new political party called the United Force. The coalition government with Brown, Jagnarine's nemesis, as Prime Minster led the colony to independence in May of 1966.

And so it was.

The fall from power of Charlie Jagnarine, former prime minister of Guyana, was considered by many an event waiting to happen. He was a friend of Fidel Castro of Cuba and Nikita Khrushchev of the Soviet Union. Because of his political ideology, he was systematically toppled in an organized coup led by a political novice named Lancelot Brown.

Years went by as Lancelot Brown quickly consolidated his power. He touted that Guyana was then a Democratic Republic, but no one believed that he was awaiting an opportunity to get on the socialist band wagon.

The PNC won elections in 1968 and 1973 despite the fact that every poll was disputed by all opposition parties, especially the PPP.

A referendum in 1978 gave the National Assembly the power to amend the constitution and elections to the assembly were postponed for over a year. The following year elections were again postponed indefinitely. In 1980, Brown seized the opportunity and, with a sudden stab of anxiety in his gut, declared himself executive President of Guyana for life. A new constitution was promulgated.

Brown was President of Guyana until he died in 1985. After his death Dennis Holmes became president until losing a close election to Charles Jagnarine.

And so it was. After nearly four decades of leadership of an opposition PPP, Charles Jagnarine had again made his way to the position of chief executive of Guyana. That time not prime minister, but president.

But his dogmatic attitude did not signal a return to socialist ideologies as in the early fifties. His furrowed brow, with the passing of time, had made him too clever to be burdened by the absurdity of socialist economics or Marxist politics. He was blissfully bereft of ideology, political philosophy and economic theory. But despite that, he nourished an existential dispute with the West, in his subconscious. But time made him a modest man. A mere Guyanese don, seizing the economic resources and political power of a country for himself and his mostly Freedom House cronies. And promoting his vision of the Guyanese national interest. Tapping the country's natural resource. Create jobs. But critics came down on him for his emphasis on natural resources projects in the lumbering and mining sectors. People thought that the concessions that were granted to the Wadpole Corporation were equivalent to giving away the country's resources. But Jagnarine felt justified in his economic initiative. He argued that decades of socialist experimentation by his predecessors had left the country without infrastructure.

"I tell you today," he told the nation during his inaugural address in 1992, "the global mining boom has begun. Guyana has no time to lose if we are going to take advantage of this opportunity. We have to think and act boldly."

But in August 1995, Jagnarine got on the national airway to announce that a gold mine owned by the Wadpole Corporation situated on the Omai River, had experienced a disaster. He told the nation that a major cyanide spill from a tailings pond in Omai had been breached spilling its contents of several billion liters of mining effluent into the Omai and subsequently the Essequibo Rivers.

"Today," Jagnarine said, his voice a lifeless monotone, "today I declare a state of emergency in the Essequibo due to an environmental disaster."

He pointed out that a 50 kilometer stretch of the Essequibo River, from the Omai mine to the mouth of the Atlantic Ocean, was an environmental disaster zone. Census figures showed that at least 25,000 people lived in the affected area. Especially Amerindians.

In contrast the Wadpole Corporation preferred to call the spill an industrial accident, much to the annoyance of the Guyanese people.

Major news media around the world stated the seriousness of the whole incident was downplayed because it happened in a poor third world country by a mine owned by a major Canadian corporation. But the chairman and CEO of Wadpole continued to maintain that there had been no human deaths and no loss of marine life as well. But several reports from Tumatumari, Mahdia and Konawaruk disputed that claim. And scientists from neighboring Brazil, Venezuela and Suriname argued that the Essequibo and its tributaries would suffer long-term damaging effects resulting from the incident.

During the next week, government and Wadpole officials toured the area and word was that they were closely engaged in limiting the damage. Wadpole was completely shut down as extensive testing of river water continued.

Chapter 4

A year after the Omai disaster, mining operations had come to a standstill. Reports stated that the mine would remain closed indefinitely. Loss to the country's export earnings, employment income and tax income was significant.

Tension was increasing between the government, environmental lobby groups and employees of the Wadpole Corporation.

Guyana's divide over the Omai cyanide spill became a heated public display one Saturday in August of 1996. Opposing demonstrators heckled and screamed at each other in the shadows of the large oak trees at Bourda Green to mark the first anniversary of the Omai spill.

Burt Samuels, leader of the GOIP, with an unexpectedly large contingent of Amerindian followers was met by smaller but multiple groups of counter-demonstrators. The counter-demonstrators were protesting their lack of work since the Omai disaster and their inability to take care of their families.

For Samuels, a stubby man of forty-five, muscular and compact as a pit bull, the protest was timed to make a statement.

"We have come here to actively express our dissent," his tobacco roughened voice sounded across Vlissengen Road through the bullhorn. "President Jagnarine is planning to reopen Omai. We, the Amerindians, are for work and the economy. But first, we must be recompensed for our loss."

Demonstrators brandished dueling placards as the marchers came within inches of each other. Georgetown police estimated that there were hundreds of Amerindians but counter-demonstrators ran in the thousands.

"Re-open the damn mines now. We want to work. Our kids are starving," the counter-demonstrators chanted.

Then others joined in.

"Yes, we want work. How can we pay the rent without work?"

Police and Guyana Defense Force soldiers maintained a heavy presence. At times keeping the opposing groups apart.

One man, claiming to represent the Wadpole Corporation stated that the company had done nothing wrong and the Ministry of Lands and Mines had its final say in all their proceedings.

"Our waste disposal methods were standard practice for the era," the man said with a resonant voice. "The Ministry of Lands and Mines approved our budget for our 1996 clean up program. It was a year after we ceased operation to clean up polluted areas."

Wadpole attorneys also disputed the Amerindians' contentions that cyanide pollution had made the Essequibo region unhealthier than other parts of Guyana. For example, "The cancer rates for the Omai area aren't different from other parts of the country," one attorney said.

Another spokesman for Wadpole said that company records showed that any health problem that plagued residents of the Omai area stemmed from poor sanitation, pesticide use and river pollution that had taken place decades ago. He told the gathering that a recent Guyana government study found illegally high concentrations of chromium chlorides, mercury and hydrocarbons in waters in the Essequibo River and its tributaries due to illegal mining operations by renegade porknockers.

The indigenous groups seeking compensation from Wadpole, however, alleged that the major cause of the region's sicknesses, diseases and deaths was the company's practice of disposing of millions of gallons of waste water from mining operations. The group pointed out that such tainted water made land infertile and rivers unfit for marine life. The group contended that the Ministry of Lands and Mines that was supposed to be our watchdog in the country's interior was paid to look the other way.

The group agreed that Wadpole and the Guyana government were partners in the Omai mining operations but Wadpole set the consortium's operating policies.

"The operation was managed by Wadpole," said Burt Samuels. "And our government gave little attention to the area except to get their cut of the extracted natural resources," Samuels yelled to the crowd, his voice harsh with indignation. "While Wadpole did use a tailing dam to sustain mining effluent, neither President Jagnarine nor Wadpole Corporation showed much concern for environmental impact," the man said. "Under Ministry of Lands and Mine laws that existed, there was no control of environmental quality," Samuels concluded.

Another man said that those who disputed the sorry state of the Essequibo region had to be out of their minds. He told the people that since the mining boom began in the

1970s, some of the area's lands and waters had been polluted and biodiverse rain forest had been chopped down. Indigenous peoples had lost land and culture. That, he said, resulted in poverty. Since the Omai disaster he continued, 85 percent of the region's residents live in poverty and lack basic education, health care and nutrition.

"We are here to vent our feelings," he continued, "these chronic situations after Omai are fueling resentment and protest against government policies. They topple presidents and threaten democratic systems. How can we live in poverty and misery while Wadpole administrators and government ministers live in inhuman wealth?" the man ended in a peremptory gesture.

During the next month President Jagnarine and his PPP controlled legislature careened closer to a fullblown legal showdown with attorneys representing indigenous Indians. A legislative subcommittee voted to authorize subpoenas for top Wadpole administrators in defiance of President Jagnarine's objections.

"After months of stonewalling, shifting stories and misleading testimony," one attorney said, "it is clear that we are still not getting the truth about the extent of the Omai disaster."

In response, an unyielding President Jagnarine threatened to rescind the authority of the legislative subcommittee that would question Wadpole's administrators, unless such interrogatories were done in private and not under oath.

Despite the rhetoric, two People's National Congress assemblymen, both members of the subcommittee, repeatedly suggested that there was room for negotiations in a confrontation that was about to split Guyana apart.

Chapter 5

The night air was thick and damp as the minibus drove south towards Seawall Road in Georgetown, Guyana. It was a little before midnight. A man seated in the rear could smell rotting sand-bitters, daisies and buttercups like a harsh perfume mixed with the heavy air of the roaring Atlantic Ocean.

Traffic was light. The city, moving reluctantly, was ready to throw in the towel.

It was August, 1996. One year after the Omai disaster. A man named Frank Wilder got out of the taxi. He paid the driver. When the taxi pulled off Wilder sat on a bench in front of the Pegasus Hotel and waited.

He waited.

He was used to passing time like that. The four year veteran of Interpol (the International Criminal Police Organization) participated in enough crime operations around the world to know how patience paid off.

His eyes swept the landscape. Georgetown was an urban terrain of dancing lights, huge buildings, tropical palms and narrow streets.

Wilder was born in Georgetown in 1958 but moved with his family to Newark, New Jersey as a youth. He didn't start out the chiseled, compact man he was.

"We used to call him 'Drum Sticks'," said Alan Radley, his high school buddy at Barringer High School in Newark. "He was a skinny little kid."

But Wilder grew up, though, knowing a bunch of heavy-lifter role models at the YM/YWCA in Newark.

The powerful image of the guys with bulging biceps stayed locked in time. It was enough to keep him jogging, pumping iron and doing free hand exercises for a long time.

Frank Wilder was deep into heavy metal by the time he graduated high school in 1976. He attended Rutgers University of New Jersey, where he pumped his six-foot-three-inch frame into a perfectly sculpted 200-pound mass of muscle.

He majored in Criminal Justice and in 1980 he graduated Magna Cum Laude. Quickly he was hired by the City of Newark Police Department, driving his patrol car or riding his Kawasaki looking for lawbreakers along Broad Street or Elizabeth Avenue.

But that day on the northern fringes of South America in a city called Georgetown in Guyana, Wilder was waiting for his contact from Interpol. After eight years with the Newark police as a street cop, Wilder remembered the day he came upon a pair of crooked cops from his department, pistol-whipping a local drug dealer and shaking down his crony. Wilder became uneasy. As the senior cop, he had to come down hard on his colleagues, especially considering the beating and failing to report a drug find, however small. Yet he also knew what was involved in one cop blowing the whistle on another cop. It was not expected to be good news when such evidence was to be passed onto the Essex County Prosecutor. Wilder looked the other way as he made a right turn with his cruiser from South Orange Avenue to South Ninth Street. He was headed to Dunkin Donuts on Central Avenue. But the whole incident still bothered him.

Within twenty-four hours, the injured bandit died. The Internal Investigative Branch of the Newark Police turned the heat on the crooked cops for their shady investigative methods. Things became more complicated when the surviving hood turned himself in, alleging that his crony was killed by two Newark cops and that he was a victim of a robbery. The hood remembered the serial number of Wilder's patrol cruiser and placed him at the scene of the incident.

Quickly Wilder hired an attorney and cut a deal with the prosecutor. He had to testify against his colleagues in the State Superior Court of New Jersey. His offense for not reporting his subordinates would be expunged from his record, providing he resigned from the police department.

Wilder grabbed the deal and walked.

But the only place where the rogue cops were to be walking was into a nine-by-nine jail cell. They each got twenty-five to life.

For days Wilder dogged at his memory, for some significant justification of his action that made him lose his job. He figured he should have nailed the rotten son-of-bitches.

Then he reckoned he had to put the whole issue behind him and move on. Because it was history.

Traditionally, ex-cops looked for work with private security firms. So Wilder thought.

All morning the following week, he worked through the Yellow Pages, trying his luck with what he had always thought of as bullshit organizations. The pay was garbage.

And his experience as a cop didn't have the spark that he depended on.

It was a handicap.

The people with whom he spoke didn't see an ex-cop guarding a trailer at a truck stop or watching expensive sports cars at a Mercedes dealer. One had to be nuts, he figured, to even consider hiring him as an executive. His experience with murderers, prostitutes, stinking winos, disheveled bums lying on park benches drinking cheap liquor wasn't a recommendation for dealing with business people.

The Yellow Pages also listed a number of detective outfits. They offered a variety of services. On inquiry, they were all one-man shows, headed by a retired police brass, not even interested in taking on an ex-cop as a side-kick.

The next four weeks, Wilder broadened his search looking for security or office work of any kind. He still got a number of rejections to his credit. Too many people were already in white and blue collar jobs, he was sorely told. And he thought of laboring? He figured he would join the laborers every morning on Freeman Street and Scotland Road in Orange for day jobs. That consisted of climbing ladders, painting, roofing or pushing a heavy wheelbarrow over bumpy surfaces, appropriate for a man his age, but not suited for a former cop with a college degree.

Then his break came one Sunday when he saw an ad in the Star Ledger for Interpol. The agency was looking for investigators, worldwide. The International Criminal Police Organization was created to assist international criminal police cooperation. He read it on the Internet. It was the world's largest international organization, after the United Nations. The United States National Central Bureau of Interpol, also known as USNCB, served as a point of contact for both American and foreign police seeking assistance in criminal investigations that extend beyond their national boundaries. Known around the world as Interpol Washington, the USNCB consulted with US police at all levels to formulate a neutral ground where states procedures were interwoven, resulting in cooperation and assistance to the fullest.

Wilder caught a NJ transit train the following day to Grand Central Station. Then he boarded the A train to Times Square and walked towards Fifth Avenue in New York City.

At 591 Fifth Avenue, he boarded an elevator in a large high rise and got out on the seventh floor.

USNCB wanted to talk to him after he expressed interest in the job with Interpol. A man by the name of Matt Kincaid wanted to see him the following day.

The next day Wilder went back to Manhattan to see Kincaid. On a mahogany colored door was written the letters USNCB. Room 703. He moved towards the door with resolute steps and opened it.

Kincaid had a waiting room that wasn't classified as classy. Drab looking wallpaper and cumbersome metal furniture. A dark green rug covered the floor. The receptionist was a strawberry blonde with fierce bullish eyes. She wore a tight fitting blue dress with slits running down the legs. For a woman in her late forties, her body was stacked like a brick shit-house. So Wilder figured. She flashed him a superior grin.

"May I help you?" the woman asked with a self-righteous voice.

"Yes, my name's Frank Wilder. Inquiring about the Interpol job."

"Sure, Mr. Wilder. Mr. Kincaid is expecting you," the receptionist continued, her expression less dubious.

She gave him an application and ushered him into a large, high ceilinged office, a place where everything that money and bad taste could have bought. Wilder sat on a chair next to a great big antique desk. Overhead, a lazy revolving paddle fan hummed as though it had seen better days.

Later Kincaid shuffled in the office. He shook Wilder's hand while introducing himself. Kincaid sported a blue shirt and yellow tie. His forbidding face exploded into an obligatory smile. When the smile vanished, Wilder was gazing on a weathered complexion face with pouchy chipmunk cheeks. Wilder reckoned that the man looked more appropriate hawking used car parts in a Wilson

Avenue junkyard. Or shoveling french fries at Blue Castle on Central Avenue in Newark rather than interviewing him for an Interpol job.

At the end of the interview, Kincaid stood up and Wilder followed suit.

Wilder noted that the man was about his size, six-foot-three-inches tall, but middle-aged and beer-bellied. Wilder thought that he himself was getting a little paunchy.

Towards the end of that September, Wilder got a letter stating that he had to complete a thorough physical examination before his application was to be considered. He was to report to a clinic at Robert Woods Johnson Medical Center in New Brunswick, New Jersey.

Upon arriving at the hospital, he was ushered into an examination room. The nurse took his blood pressure, pulse and respiration and drew blood. A doctor with hollowed cheekbones laid him on an examining table and checked him out from toes to scalp. The man's name was Dr. Hoffman. He was one of those old school guys who walked around puffed up with self-importance. He wore a bow tie and cuff links under his lab coat. Although there was a sign on his wall that said 'Smoke free environment,' the place reeked of cigars.

"How do you feel, young guy?" the man croaked.

"Fit as a grizzly," Wilder remembered answering.

"What are you——twenty-seven, twenty-eight?"

"Thirty."

"Splendid. You're in good shape," the doctor said, his rheumy eyes looking over his glasses.

Two months later, Frank Wilder was hired by Interpol's General Secretariat, joining a staff of nearly four hundred police officers and civilians. That same month, the agency changed from a nine-to-five agency to a twenty-four-seven organization, making it easier and more efficient to operate in fifty-nine different countries.

Interpol's Constitution mandated that it did not intervene in crimes that did not overlap the borders of member countries. Nor involve itself into political, military, religious or social crimes. Its concerns were hinged on public safety, terrorism, organized crime, illicit drug production and trafficking, weapons smuggling, human trafficking, financial and high-tech crime and corruption.

Wilder worked for Interpol for the next eight years. His first three years were spent in California, New York and New Jersey. Then he was sent overseas to Rwanda, Togo and the Ivory Coast for two years. Then Wilder came back to the US. After being in New Jersey for three years he received a letter from Interpol notifying him that he was assigned to Georgetown, Guyana.

Chapter 6

Frank Wilder hung out by the Pegasus Hotel for about thirty minutes. Suddenly a blue Ford Bronco pulled up and parked next to a Volvo about two hundred yards from where he sat. Wilder heard the door of the SUV open and close. Not far down the cinder sidewalk, he saw a man as stocky and compact as a tapir, moving towards him with a proud, military gait. His body loomed steadily, unwavering and seemingly grotesque, as something seen through a jaundiced eye.

As the man drew closer, Wilder noted that he was in his mid to late forties with a fleshy moon-shaped face. His mouth set in a permanent frown.

"You must be Frank Wilder?" the man's husky voice sounded as he stretched out his hand with claw-like fingers.

"Sure, you must be Fred Nelson." Wilder said with a slow appraising glance.

"That's right. I'm Fred Nelson, chief of Interpol in Guyana, stationed at Eve Leary."

"Yep. I know Eve Leary."

"You're not kidding. Records show you were born in Guyana."

"Sure. Left Guyana for New Jersey when I was young."

"Man, you were gone a long time."

"Yes. Twenty-five years."

"That's a long time. Still remember places?"

"Time will tell, I guess. Haven't been around since I landed at Cheddi Jagan's Airport two days ago. Who knows. I may or may not have the time to check out the old neighborhood in Alberttown."

"You're right, Frank. Won't be too many places to go. Interpol is now a 24/7 job and here in Guyana, the shit is about to hit the fan."

"No kidding," Frank answered, his voice unflinching.

Nelson noted the grittiness in Wilder's voice but continued.

"Look it here, man," his sardonic voice piercing the night air, "I'm from Lancaster, Pennsylvania, been with Interpol for twenty-two years. Was in Haiti, Antigua, Zaire and Portugal. But here in Guyana, man, it's a different story. You're talking about corruption to environmental problems and more."

"No job is a walk in the park, I guess," Frank added.

"You're right about that Frank. You too by the way got some experience in foreign countries."

"Yes, I've been in several of them."

"You'll need that experience here."

"So what's going on now?"

"It was a month ago that President Jagnarine reopened the Omai gold mining operations but GOIP and its leaders said 'no way'."

"So what went down?"

"GOIP and Burt Samuels, one of its leaders, were celebrating Amerindians Heritage Month to coincide with the mining incident one day on Vlissengen Road, not far from Jagnarine's residence. About an hour after their gathering, this pick-up came on the scene with four gunmen all with bandanas. They started shooting."

"Anybody got hurt?"

"You can bet on that."

"How many?"

"Burt Samuels and six members of GOIP were killed."

"Holy mackerel. That's enough to start some problem based on what I read about the cyanide spill at Omai."

"That's right."

Nelson explained that he was sent to Guyana in 1994, a year before the Omai mining disaster. He stated that the reason for him being sent to Georgetown by Interpol was in response to allegations of human rights abuse. He claimed that the allegation was determined by a past United States president who visited Guyana in the late eighties and early nineties and found that the electoral process was unfair.

But unrest continued during the months following 1994, Nelson said. And so were the violent crackdowns. Charles Jagnarine, stubborn and unrepentant as ever, had vowed to come down hard on antigovernment protestors since 1995. He dismissed international criticism of his handling of the Omai mining disaster as an imperialist plot.

Although antigovernment sentiments were triggered by the administration's lack of respect for human and political rights, Jagnarine's poor management of the economy was also to blame. Guyana's inflation rate exceeded five hundred percent and is considered to be among some of the world's highest.

But Guyana's cyanide crisis, according to Nelson, raised familiar questions about the responsibilities of the international community. Jagnarine argued that the world had no business interfering with, or even commenting on, the internal affairs of a sovereign state. That principle, held by Jagnarine, was exceptionally convenient for dictators and people who did not care about the well-being of others.

"It is a mind set that paved the way for the rise of Adolf Hitler, Josef Stalin, Pol Pot and Idi Amin!" Nelson said.

"That's right," Wilder nodded in agreement.

Nelson said that after Burt Samuels was murdered, his lieutenant, Melvin Richards, fearing for his life fled to neighboring Port-of-Spain in Trinidad. There in a city of over one million people, Richards got on TV and suggested that the world should intervene to bring closure to the Omai disaster. Richards said that global and regional organizations and individual governments should make known their support for human rights and protection in Guyana as elsewhere.

"People should condemn in the strongest terms the use of violence as a way of replacing due process for the Amerindian people," Richards screamed, with grave deliberation. He told the commonwealth nations that Charles Jagnarine had showed a consistent unwillingness to respect the legitimate complaints of the indigenous peoples of Guyana. Richards called on the United States to show concern. The time for silent diplomacy was over. People had to speak out, then or never.

Wilder took a cigarette from his pocket as Nelson finished his sentence. He stuck the Newport between his lips and struck a match. He inhaled long and deep like a stevedore. When he exhaled, his hand fell to his side in a characteristic gesture.

"This guy, Richards, sounds like a lightning rod," Wilder said.

"You're sure right. He was educated in Trinidad and was the only man who had the balls to lead GOIP after Burt Samuels and others were gunned down. For decades there was an inherited ideology in the minds of Guyana leaders of not to consult with the Amerindians when woods were cut on their land, when mining, forestry and oil exploration concessions were negotiated with outsiders like owners of the Wadpole corporation."

Richards told the people of Trinidad and Tobago during his thirty-five minutes interview with Baldwin Lindsay of TBTV channel six of Port-of-Spain that Charles Jagnarine had met his match in Burt Samuels. He said Mr. Samuels intended to give the indigenous community the opportunity to forge a bond among the nine tribes of Guyana so that they would be able to show solidarity with their colleagues in their battle against Mr. Jagnarine and the Wadpole Corporation. But, he said the government of Guyana had its plans when Mr. Samuels was gunned down in Georgetown in the midst of a peaceful gathering. But Melvin Richards said he wrote letters to Charles Jagnarine through the Embassy of Trinidad and Tobago pledging his continued demonstrations and civic disruption at the Guyana Embassy in Port-of-Spain if there was no independent inquiry into the Omai disaster and the death in Georgetown of GOIP leader Burt Samuels and six of his followers.

But Richards' entreaties fell on deaf ears.

"That guy Richards is real gutsy," Wilder said after a long pause.

"Yes…you're right," Nelson answered.

"So what's next?" Wilder appeared puzzled.

"That's where you come in," Nelson responded as his face hardened into a serious expression. "You've got your orders. Get some sleep and move on in the morning."

Chapter 7

The dining room at the Native Restaurant on Sidney Street in East LaPenitance was more like a cook-shop in Stabroek Market. The place was noisy, frenzied and boiling hot. Ginger beer and sorrel drinks were being handed out. Wilder grabbed one from the passing waitress as everyone else seemed to do. The place was packed. Wilder saw no vacant chairs. But then, a man who Fred Nelson had described, waved him across the room. The man with a lanky body pointed to a chair next to another man.

"So," he said, his voice energizing, "you made it to the Native."

"Yes, sir," Wilder said. "You must be Martin Dubbin."

"Sure, I'm from the US EPA. This is Keith Galloway from the World Health Organization," Dubbin said, pointing to the man seated next to him.

Galloway had a low forehead. His lips, a thin horizontal line.

"What are you eating?" Dubbin asked.

"Guess I'll try the vegetables, mashed potatoes and stewed tilapia," Wilder answered.

"Keith wants fried chicken but I'll go for a T-bone steak," Dubbin added.

"Now what's with Omai?" Wilder asked.

"Hundreds of people died from cyanide poisoning after the dam broke," Dubbin said. "Somebody in authority within the government is covering up those deaths."

"Is that right?" Wilder asked.

"Sure it's right," Galloway added. "All along the Essequibo River tributaries like the Potaro and Burro Burro Rivers, people have been drinking river water as they always do. They either died from cyanide poisoning or leptospirosis."

"What's that?" Wilder asked.

"Exposure to water contaminated by dead animals after the disaster," Galloway answered.

"So what's the point in keeping the truth from the public?" Wilder asked, breaking his few moments of silence.

"Economics, man," Dubbin answered.

"That's right," said Galloway. "Investment in Omai as of now totals about $800 million. And that's US dollars now."

"No kidding," Wilder answered.

"According to reports, the mine produced more than a quarter of a million ounces of gold last year," Dubbin added.

"What on earth Wadpole did with all that gold?" Wilder asked as he regarded the men curiously.

"That stuff was sent to Canada for refining," Dubbin said.

"What did it do for the Guyanese people?" Wilder asked.

"The economic contribution to the country was beyond belief," Galloway said. "Just personal income taxes paid by mine employees represented sixteen percent of income taxes collected throughout the country."

"That still doesn't include money paid to government officials to look the other way," Dubbin said.

"That's shocking," Wilder said.

"One didn't have to wonder why President Jagnarine rescinded the authority of the legislative subcommittee that was formed to question Wadpole's administrators about their business practices before, during and after the Omai disaster," Dubbin said as his eyes widened in alarm.

"What's more shocking, guys, is that the Ministry of Health has long since been dragging its feet on disclosure of morbidity and mortality statistics in the Essequibo region after Omai," Galloway said as he took a huge gulp of his sorrel drink.

"Getting to the bottom of this whole issue will take some time," Wilder said as a stoic expression crossed his face.

"That's right. It will take time. How much? I don't know," Dubbin said rhetorically.

"But you may want to know that the mine employs over fifteen hundred people," Galloway added.

"That's a lot of people," Wilder said.

"Shouldn't be a problem for you, Frank. You've worked as a cop in a huge metropolitan area for years. You've seen the worse," Dubbin said.

"I ain't complaining," Wilder answered.

Chapter 8

Frank Wilder made his way towards the Margarita Hotel on a warm November evening along Middleton Street in Georgetown. He was wearing baggy jeans, a blue T-shirt and black sneakers. A pair of mirrored aviator shades hung over his eyes. His gait was athletic, but not elaborate.

At that time the rush hour was omnipresent, compared to when he left Guyana at twelve. The city was growing at an alarming rate.

He opened the door and approached the clerk. She was an Indo-Guyanese about middle age.

"Have you any rooms?"

"Yes. What kind are you looking for?"

"Single room. Queen size bed."

"One's available on the second floor. Room 203. Here's the key. You can check it out."

"No need for that. I'll take it."

"Fill this card," the clerk said, sliding it across the desk. "It will be five hundred dollars a night. How long are you staying?"

"Indefinitely. I guess until I find an apartment in some part of town."

Wilder grabbed his suitcase and shuffled towards his room. He was tired. He figured he should be. All the running around after leaving New Jersey just four days ago. Who wouldn't be spent? He pondered.

The room was absurdly small. Cheap wallpaper decorated the walls. While two grimy windows, one with a cracked pane, gave some light to the surrounding. The carpet had once been white or off-white but constant treading had reduced it to a light brownish hue. Two incandescent bulbs hanging from an outdated fixture adorned the center of the ceiling. Time had shown no mercy on a gray leather sofa that lay against the far wall. The bed was nothing more than a piece of plywood on four legs.

Wilder had a stricken look on his face as he entered the room. It was Fred Nelson's idea, Wilder remembered. The plan was for him to leave the ritzy Pegasus and move to a less upscale part of the city, blend with the denizens and find the Omai files.

He sat on the sofa, arched his body backwards and pressed his head against his interlaced fingers and thought of Sofia McKean. Was it two or three years ago, he thought. Time sure flew. She was a respiratory technician at St. Barnabas Medical Center in West Orange New Jersey. And his woman. One day he got word from a buddy that his girl was fooling around with some young emergency room doctor. Wilder didn't know what to make of it. But he never confronted Sofia because there was also a possibility that the whole story was nothing but a rumor. Just a demon in his head. Other times he wanted to think the worst. He didn't know why. It was as though something deep within his brain required the certainty of a confirmed betrayal. That confirmation came one Friday evening at the Peppermint Lounge in Orange. She was in the arms of another man. But maintained he was just a friend. Wilder kept his composure by efforts that brought the sweat to his brow. He remembered the sudden spurt of adrenaline coursing through his veins when his voice exploded, "It's over, goddamn it. It's over!"

But as he sat on the sofa at the Margarita Hotel, the memory of the event haunted him. He fought with his feelings, for justification of his action. Her eyes, he remembered, once moist with joy, had a deepened hue of shame. It was sad, he thought, the way it ended. He felt jaded as those memories flooded his dulled mind.

Chapter 9

Edward Hartman was a big beefy training instructor with a face puffed up with constant inner turmoil. As a member of the Guyana Police Force, he had been training recruits for decades at the Police Training College in Thomas Lands in Georgetown. That day he stood before a platoon of men. They had already passed their preliminary test and must undergo a rigorous eight week course in police work.

Each man was called a constable and upon graduation, he was to become a member of the security team at Omai gold mine.

Frank Wilder was one of those men.

The beefy man, well over six feet, eyed the men before him with skepticism. He weighed each man's facial expression with a critical squint.

"I'm loathsome to look at. In other words, I'm an ugly son-of-a-bitch!" Hartman said, spitting out his words with contempt. "But I don't give a monkey's ass!"

His khaki uniform was immaculately clean. His boots mirrored against the morning sun. He took a circuitous path toward the platoon, his baton under his left arm held parallel to the ground.

One constable in the front row inadvertently made contact with Hartman's eyes. He stormed towards the recruit who was about an inch below Hartman's height. He peered down the man's nose and with his face hardening with anger, his vociferous voice sounded.

"Don't ever! Never ever look at me straight in the eyes! You poor scum. You don't deserve my gaze. When you look at me, make sure you use your peripheral vision!"

The recruit didn't respond.

"Did you hear me constable?" Hartman roared.

"Yes…sir," the recruit acknowledged.

Hartman moved with a proud military gait and again stood in front of the platoon.

"In this business we don't hire mamas boys or pricks. We hire men. What you're here for is not a walk through the park. You'll be guarding and escorting gold shipments from the Omai mines. At times you'll be going after scumbag porknockers, thieves who infiltrate the mine's perimeter to steal reclaimed carbon that contains traces of gold. Don't get me wrong. Not all porknockers are thieves!" As Hartman ended his sentence, a smile of malicious delight crossed his face. Then he continued. "You men, after graduation, will be joining a security force of over one hundred men. Your boss is the superintendent of the Guyana National Police Force. He reports to the Commissioner of Police. Are there any questions?" the man asked.

No one uttered a word.

"Good!" Hartman taunted.

But for Frank Wilder, the scene brought him back to the days when he was a recruit at the Essex County Police

Academy in Cedar Grove, New Jersey. And déjà vu all over again.

It was his weapons instructor at the academy. An alcoholic and an abusive cop. Verbal abuse that was. He swore at everyone for no reason. But Wilder drew strength from his rough family upbringing in dealing with such personalities.

His mother was never in his life after he came to the United States with his family. Report had it that she just disappeared somewhere in Times Square, Manhattan, one day on her way to work. As a consequence, his father resorted to drinking and was a full blown alcoholic by the time Wilder turned fifteen. Then he remembered the last time he saw his father alive. The older Wilder staggered to his son in an alcoholic stupor and hugged the younger man. Frank smelled the whiskey all over his father's body. Like it was coming out his pores. Frank's face was wet from his father's tears and his spit. He was unable to speak as his father cried all over his head.

That night he died.

Next day Frank and his young sister were taken by the New Jersey Department of Youth and Family services over alleged abuse and neglect by their deceased father.

Years later, Frank told a friend at the academy that his real-life experience had taught him how to compete through adversity. No matter what the situation was.

And he never did forget his guardians on Roseville Avenue in Newark. The Plummers. Good people. With a few kids of their own. They didn't adopt Frank and his sister but sort of took them in for years. They went with the family to Roseville Presbyterian Church and attended Barringer High School with the Plummer's kids, Keith, Gerald and Audry.

That part of Newark was a tough environment with all the different ethnic groups, Frank remembered. Mean, hard-boiled bastards. Tough as nails. As a teenager, he saw a lot of so-called scrappers. He was one himself. At times as mean as a junkyard dog. But yet in all his life he didn't think he had ever met a so-called bruiser who couldn't in some way be brought to his knees like the guy Hartman. What a stupid obnoxious sap he was, like one of those Monday morning quarterbacks. A man who thought he was bad. Maybe one had to dig deep or hack away at his psyche to expose his underbelly. Because Wilder believed that everyone had an Achilles heel. And in Wilder's mind, he knew he was prepared to handle guys like Hartman with his macho posturing and scathing insults.

Then he remembered that day in New Jersey when his instructor uttered language directed at him that he found insulting. Wilder gave him a scorching look. The man, well over six feet with wrestler's arms and iron stomach muscles, noticed Wilder's anger. He goaded Wilder to step before the group and meet him face to face since he was so pissed off. Wilder told the man that he respected his rank and age but he had issues with his uttered words. But those, he said, were not justifiable reason for a confrontation.

"To hell with your feelings!" Wilder remembered the man screaming in exasperation. "If you don't bring your ass here, I'll come and get you!" The instructor continued as an edge of impatience crept in his voice. "And remember this is off the record!"

Wilder's face grew ashen with bitter indignation. He swallowed hard on a bolus of saliva as he approached the instructor.

Before long, only eighteen inches separated the two.

Wilder looked in the man's face. It was gaunt for his age at forty-one. The eyes were bloodshot and the cheekbones were high and ruddy. His nose projected from his face like a cucumber. Then he flashed Wilder that dirty gap-toothed grin and clobbered him with a body shot, right to the solar plexus. Wilder crashed to the ground clutching his stomach. A small groan of agony hissing through his teeth. Three years prior, at nineteen in Jersey City, a scrappers' fist had broken his jaw. Two years and two surgeries later, it had finally healed.

"Get up sucker. On your feet!" the man screamed at Wilder with a mocking grin while the other recruits looked on in awe.

Wilder regained his feet with a cat-like grace.

"Don't hit me again man!" Wilder warned the man, as he went back a step.

"By Christ, I'll show you that this is the Police Academy of Essex County and assholes like you will respect guys like me!" the instructor shouted. There was a note of disbelief in his voice. To his knowledge it was the first time a recruit had challenged him. He looked at the younger man in the eyes and his anger boiled more violently than before. He took another swing at Wilder. But Wilder twisted away. But not fast enough to avoid the tip of the man's knuckles that grazed his shoulder. Again the man cranked up his body and with his fists bunched, he took another shot at the young recruit. Wilder ducked the blow and pushing his arms backwards for balance, he delivered a stunning front kick to the man's mid-section. The instructor reeled backward with his face contorted in agony. Before he hit the ground, Wilder was on him. He punched the man in the middle of his exposed chest, putting the full weight of his body into the blows.

Quickly the other recruits dragged Wilder off the man. "That's enough man. You'll kill him!" one man shouted.

The instructor regained his feet. A snarl of pain blanketed his face. His lips were drawn tight, baring his teeth. His eyes no longer blue, but burning black from inner pain. The man sucked in big breaths of air. His energy was coming back.

He walked over to Wilder, struggling to control a quavering voice. He grabbed the young recruit by the hand and shook it. When his trembling lips parted, he whined, "You'll be a fucking good cop, my boy."

Wilder nodded in approval.

Chapter 10

Like a giant anaconda, the Ayanganna Highway snaked its way through Linden then west towards Rockstone. It meandered across the Essequibo River then south to the Omai mine. The site occupied a hefty chunk of real estate in the Essequibo region. Thirty-five square miles of land lay sprawling on the West Bank of the massive Essequibo River, encircling the Omai River. The scene was like out of a James Bond movie. Trucks, trailers, forklifts, bulldozers, excavators, backhoes and shuttle cars dotted the landscape. While uniformed security forces, some in personnel carriers, others on foot, brash and forbidding, criss-crossed the area. They all packed heavy fire power. Some carried AK-47 assault rifles, M60 machine guns, Uzi submachine guns and Mossberg shotguns. Smaller arms like snub-nosed .38s, Browning high powered 9 mm and Smith and Wesson .357 completed their arsenal.

Five hours had gone by since Frank Wilder and the other recruits had left Georgetown. For them, the first part of their training at the National Police Training College was completed.

The final part of their basic indoctrination was at hand. It was to take place in the mine's environment.

It covered information relating to mining operations, security at Wadpole's headquarters in Georgetown, the wharf at Linden that was the company's supplies and equipment depot and last but not least, the mine site at Omai. Security officers also known as ranks or constables, had to be knowledgeable in the techniques of hand to hand combat or to have the ability to disarm an adversary, brandishing a club, machete or a cutlass.

It was 10:30 a.m. when the big Bedford, a 3-tonner truck, rolled along a dusty Omai road. It passed multiple pits and large wooden buildings designed for processing ore. There was also a big purification plant, warehouses and a power station. Smaller structures dotted the landscape. Workers were everywhere, performing their tasks with diligence.

Surrounding the mine, the jungle stood out like a green enclosure against the morning sun. Giant, graceful mahogany and greenheart, cedar and wallaba and palm trees towered, almost in the face of heaven. And not far beyond, to the south, slithered the Omai River. Devoid of rest. Tumbling and hissing with a constant moving sound.

The silence of the jungle despite the belching, spitting, hawking and vomiting of the heavy machineries, meant something to Frank Wilder. Its mystery. Its magnificence. The perplexing reality of its concealed life. To him it was an enigma until that night in August, 1995 when the dam broke and left behind a trail of death, disease and destruction. But that day, deep in the Essequibo jungle, Wilder had to come to grips with the reality of his mission. Finding the Omai files and exposing a cover-up that involved the lives of hundreds of innocent people. The philosophical aspect of his musings had then to take the back burner. He had to accomplish his mission. But could he handle the dumb thing, or would it handle him? He wondered.

Ten minutes later, the truck pulled up along an administrative building. The recruits, numbering twenty-four, filed out towards a lecture hall. Inside, an Indo-Guyanese man about fifty-four stood by a lectern. A police sergeant who had accompanied the recruits from Georgetown to Omai sprang to attention. And so did Wilder and the other recruits.

"At ease, men," the man said after returning the sergeant's salute. "You may take your seats. You too, sergeant."

"Thank you, Superintendent Balkarran," the sergeant said in a voice overly polite.

The man with a deep scar on his right cheek was Roger Balkarran, Assistant Superintendent of the Guyana National Police Force. The flesh of his timeworn face was taut, as if held in place by glue. The effect was neanderthal, macabre. When he spoke, his voice grew to a maniacal crescendo. It heightened the image of a man not totally well, as if being eaten on the inside by termites.

"Welcome to the Guyana National Police Force. You have chosen a difficult job," the man said with a heavy gravel voice. "Those of you who are not serious of being police officers, you can back out now, with no questions asked. If not, be prepared to go the full nine yards. Any questions?" Balkarran asked.

"Are we allowed to use deadly force in times of severe threat?" one recruit asked with a tight frightened voice.

"You follow the procedure of the Guyana National Police, my boy. If I'm chasing someone down the street who's brandishing a weapon, make no mistake——he's out to kill me. Still, I'm supposed to follow procedures and use my judgment to make sure no one gets hurt. But if it's me or him, better him than me. We can't let criminals and law-breakers run over us. We take them out when necessary."

During the next week, the recruits were inundated with classroom instructions relating to police work. The last three weeks found them involved in weapons training, heavy physical exercises and martial arts for defense and offense. They also completed five days of marine and river operations, jungle survival techniques, map interpretation and emergency medical training.

For most of the recruits, those final days of their training were etched in their minds and would never be erased.

For Wilder, it was reliving those agonizing days at the Essex County Police Academy. At times after training, others would say to him, "Man you made that look so easy." But no one in Guyana had the slightest hint that he was a former Newark cop and at that moment an Interpol investigator. Except, of course, Fred Nelson, his boss at Eve Leary. Also Martin Dubbin of the US EPA and Keith Galloway from the World Health Organization.

That day it wasn't like he was twenty-two and a rookie cop. Besides his New Jersey training was a lot different. There was no jungle. Presently, he was at the back end of his thirties. And here he was determined to keep up with a bunch of younger men in his squad. The ordeal took its toll. Not only on his body. But at that moment his overall training at the Omai site rattled him to his psychological core.

Nevertheless, he welcomed the challenge with his mind and body. It taught him to maintain composure and deliberateness in the midst of physical and emotional mayhem.

Chapter 11

The wildcatters were out on their own. Their only hope was to find that piece of gold that would guarantee them their most elusive dreams. A mansion in a Georgetown suburb. Servants and a big fancy car. Or a vacation in Barbados.

Sometimes they are known as porknockers or freelance prospectors who after exhausting their own claims saw the entire Essequibo rainforest as theirs for the taking, despite demarcation zones. Those adventurers operated by force or fury. By stealth or by sting.

The indications of their encroachment on the Omai mining property's fenced compound that night was very clear. Fences were cut through by a heavy wire cutter. Obscure manmade trails on the jungle floor. The crooks had hauled off large amounts of reclaimed carbon from the mine. The perpetrators were savvy. They were aware that carbon was utilized as a means of removing gold from the ore. And the residual carbon had particles of gold.

Roger Balkarran, the assistant superintendent of police in charge of security at the Omai mine, got a memo concerning the theft of reclaimed carbon from the lieutenant in his charge. The assistant superintendent's face wore an implacable expression as he summoned the lieutenant to his office immediately.

"What the heck is this, lieutenant?" the assistant superintendent blurted, making a disgusted gesture as he slid the memo across the polished desk in the direction of the man standing before him.

Tension caused the lieutenant's voice to crack as he shifted and reshifted his feet.

"It was last night...sir. The porknockers struck again!"

"Where in hell were the guards?"

"On duty, sir."

"Then what in hell happened? This is serious, you prick. We can be reassigned if this continues or even get fired!"

"Be realistic, sir. Staying on top of those porknockers isn't the easiest thing in the world."

"Oh yea! You got some balls, lieutenant. Like hell, that doesn't mean we must let those shit heads cut through the fence and steal whatever they want."

"No, sir. It's a problem of geography. We're talkin' about a site as big as fifty square miles in area."

"Ok egghead. Take me back to school."

"Jungle on all sides except the east that has the Essequibo River."

"Don't you think I know that, lieutenant. What's your point? You're wasting my time. Damnit!"

"Sir, I believe that the porknockers are making their way along the Essequibo River, deep in the bayou region, southeast of Tumatumari."

"You think so," the superintendent responded with some decreased hostility in his voice.

"Yes, sir. We've got armed guards along every point of mine property. But the experienced thieves are watching. A hole in the fence was a diversion. They navigate those piranha- and reptile-infested creeks leading to the mine from the Essequibo River."

"I hope you're right, lieutenant. For your own bloody sake."

"They must have a boat in their possession and move under cover of dark to elude our marine and river units."

"So what's your plan?"

"Get a few good men to follow their trail. They couldn't be too far especially having to carry their heavy spoils. They'll be arrested when caught."

"Lieutenant, how long had you been a cop?"

"About seventeen years, sir."

"Most of that time you've spent in Georgetown, right. Going after criminal elements."

"Yes, sir. Been at Omai for three years."

"Well, lem'me say this, lieutenant. I've been here at Omai in charge of security for damn near twenty-two years. And I never in years past, allow a thief to escape me. No. Never."

"I'll do my best, sir," the lieutenant said in a strained voice.

"Your best is not good enough, Lieutenant. You better plan hard. Work hard and hit hard. Remember, in this damn job there's no room for error. No room for compromise. Hit those suckers like a bolt of lightning. Bring 'em in now!"

"Sure, sir."

"So, get the ball rollin'. Wait, by the way. You've got some fresh ranks, I see."

"Yes, sir."

"Think they're ready for this job?"

"Two are, perhaps three."

"Is that right?"

"Yes. Two are former Guyana Defense Force soldiers. They know the jungle."

"And who's the other man?"

"His name's Frank Wilder. Left Guyana at twelve to live in the USA. But he came back after being there for about twenty-five years. He didn't give a reason. Guess he got homesick."

"Can't blame the fella for missing his home, if that's the case. Is he any good?"

"Yes, sir. Training wing said this man exceeded expectations in all aspects of his training. Police work, combat, weapons skills and jungle exploration. He's got it all."

"There you go. Utilize his skills. You got your squad in place. No bullshit now. Get to hell outa' here, Lieutenant!"

"Sure, sure thing, sir."

Chapter 12

Sergeant Phillip Hyman and Corporal Cyril Kumar, both former Guyana Defense Force soldiers were among ranks who had left the army to join Omai Security Force. They had been trained to protect Wadpole's concerns with equal verve.

The former soldiers pursued their love of the armed force in a second, but that time, paramilitary career at Omai.

After intensive jungle operations, the pair was ready for high-profile missions.

Hyman, born in Rose Hall on the Courentyne coast was the grandson of a former major of the Guyana Volunteer Force, disbanded some four decades ago.

Phillip Hyman dreamed of being a member of that kind of organization since he was a little boy. His Uncle Jeff would take him to see war movies on weekends. Or at times he played war games with plastic guns around the bushes with neighborhood kids. That day in Omai he wore with pride the chevrons that depicted his achievement with the Guyana Defense Force and the Omai Security Force.

His military career began when he joined the Guyana National Service Program at eighteen. It was a three year mandatory program for all Guyanese. When his term expired, he left Thomas Lands, Georgetown to begin his journey on a sixteen week ordeal with the Guyana Defense Force at Camp Ayanganna. He received the army's expert marksman shield for his one-inch grouping on the rifle range. And high marks in small arms and self-defense skills.

Cyril Kumar, on the other hand, born in Georgetown, descended from a long line of Indo-Guyanese merchants who controlled most of the textile industries lining Regent and Robb Street in the country's capital. His father sent him overseas to study political science at the University of Cambridge in England. But he returned to Georgetown a year later, claiming that he wanted to pursue a military career.

It was at the Omai mine where the two men crossed paths with Frank Wilder.

Hyman was a lanky man of thirty-seven, with a defiant smile. And Cyril Kumar, a medium built man of thirty-five, not stocky, but carried a tightly pinched expression on his face.

Hyman was the leader of the trio.

"Ok, men," he said pointing to a red dot on a map, showing the Essequibo region. "The copter will drop us at this point. We're well aware that the porknockers have a six-hour start ahead. But we've got to hit 'em before they reach the Essequibo River."

"What if they slip through our dragnet?" Wilder asked.

"Even if they have a boat, escaping us will not be possible. The copter will put us miles ahead of them. They'll come in our direction."

"Besides we have river patrols guarding the mine property that borders the Essequibo River," Kumar added.

"That's right," Hyman continued, "but we've got to hit 'em before they reach the river. Use your cell phone when necessary. It's your life line. Any questions?"

Both Kumar and Wilder's response was negative.

"Good," Hyman said. "When we get there and start our move, we gotta split up in three. Cyril will take the left flank. Frank to the right and I'll bring up the middle."

"You fellas got it?"

Both men nodded their agreement.

"Ok. You got your orders, but adjust to situations as they arise. Remember these men out there are vicious. They're full of surprises. So things can get ugly. But keep your wits about you and stay in overdrive."

Then the drone of a helicopter was heard. Simultaneously, a green Toyota Land Cruiser rolled up on Starapple Street, site of the administrative building at Omai. The lieutenant emerged from the SUV and approached the building with mincing steps. He hurled himself through the door and met the three men in a conference room. The lieutenant's face grew haggard with worry when he spoke.

"The chopper's here, men," he said in a tight voice.

"Ok, sir," Sergeant Hyman responded then, turning to Kumar and Wilder, he added, "It's showtime, guys. Let's move."

"Do what you've got to do, fellas," the lieutenant added cracking his knuckles with anxiety, tension and nervousness.

An hour later, deep in the devilish heat of the Essequibo jungle, three men were on the move.

They stuck to their plan. Wilder to the right. Kumar to the left while Sergeant Hyman brought up the middle.

The men disappeared in the dense undergrowth.

Wilder charged along the jungle floor, making his own path as he burrowed into the thick foliage with machete in hand.

One hour went by.

For Wilder, the vegetation was particularly unforgiving. Layers and layers of green canopy ran overhead. Greenheart, purple heart and wallaba trees rose from buttress roots to heights of almost four hundred feet. Below them smaller trees like cedar, bloodwood and bullet-wood trees predominate. They produced a second canopy. It was thick. No light was able to reach the jungle floor. There seedlings struggled to encounter light.

Wilder heard a momentary hissing. He stopped in his tracks as his eyes peeled about the jungle floor. He quickly grappled a mass of vine and lianas and pulled himself to safety on a large moza root. Ferns, mosses and herbaceous plants ventured through a thick carpet of moist vegetation. There among fallen tree trunks and rotting leaves, Wilder spotted a giant boa constrictor. The reptile's blood-red eyes held him with a murderous gaze. Its opened mouth was frothy. Its black tongue slithered in and out of its mouth.

The giant snake did not move. It didn't take Wilder long to figure out the reason. The reptile's midsection was bulky. It had just swallowed its meal. Might have been about two hours ago. Wilder reckoned that an anteater, a monkey, deer or tapir might have fallen victim to the reptile.

He continued his push towards his quarry.

Ninety minutes later, he gazed at his translucent watch. It was almost 5 p.m. He stood at the base of a large mahogany tree, thinking of a safe place to pass the night. Then the sound of primates engulfed the boughs of the big mahogany tree.

Wilder's eyes widened as he gazed overhead. Night monkeys with large eyes and varying coats, followed by howler monkeys, spider monkeys and capuchin monkeys, spilled from the branches. They made acrobatic leaps to the safety of neighboring trees because of Wilder's presence.

He nevertheless, ascended the tree. At its highest point, Wilder noted that the sun was dwindling to a distant meeting with a pale horizon. Before it, lurked purple, orange and burnt-almond clouds. As if gathered in a last magnificent trilogy before darkness fell from the bosom of night.

Chapter 13

It might have been just after 7 p.m. High in the mahogany boughs, between a set of three-pronged branches, Frank Wilder wedged himself.

He intended to find some repose even if it meant he had to engage in cat naps. Despite the insect repellent he smeared on the exposed areas of his body, the mosquitoes kept up their attack with reckless abandon. For Wilder, the vectors persistence had reached the very zenith of annoyance.

The insects were unable to settle on his body. Thanks to the repellent. But in the dark, they buzzed by his ears, nose and eyes. Once in a while, he felt the furious stings of the determined creatures on his back as their proboscides penetrated his shirt and into his flesh.

But the mosquitoes weren't his only nemesis.

Wilder thought he might have nodded for about ten minutes. But he was rudely awakened by the brief sound of flapping wing before his face. It caused him to panic momentarily. But he tried to compose himself as he sat in the dark. Straining his eyes to locate the object of his concern.

Again the sound returned, like flapping wings before his eyes and then vanished.

Quickly he whipped out his cigarette lighter and when the sound returned, he struck the lighter.

And there for a brief moment before his eyes, Wilder espied the creature. A large vampire bat, with at least a three feet wing span. Long, sharp teeth decorated its mouth. Rumor had it that the creature was considered the blood-sucking ghost of South America.

But Wilder did not believe in rumors. He was too practical for that.

Only the thought of being bitten sent a rush of heat throughout his anatomy.

Although that was his thought, he remained undaunted. Quickly he hacked off a small branch nearby and began swinging it wildly in all directions in an effort to scare away the bat. He continued that for five minutes until his arms grew tired.

Then he sat quietly in the darkness.

But within seconds, he recoiled in horror at the sudden outburst of a group of howler monkeys. Wilder reckoned that the noise he made to scare away the bat might have startled the primates.

The howler monkeys bellowed and bellowed. Wilder sensed their hard-driving hostility. The blasts were deafening as they echoed through the jungle and beyond.

After twenty minutes, the noise ceased.

Wilder spent the next several hours sleepless in the mahogany boughs.

Then the beckoning cries of macaws, toucans, parrots, eagles and hawks heralded the break of dawn.

Wilder hurried down the tree at sunrise. The fog was heavy everywhere. He stood in one spot for a few moments, looking around him and listening. The only sound he heard was the growling of his stomach. Or the occasional cry of a tapir or jaguar.

Then he wondered whether Hyman and Kumar had spent a miserable night like he did. He continued his push in a northeastern direction.

Two hours went by.

The humidity of the jungle was increasing. Then deep in his ear, he knew he heard a sound. Like some huge animal, moving on dry underbush. He plastered his body along the trunk of a large bullet-wood tree. And waited.

It wasn't long before a giant tapir came into view. The animal weighed about five hundred pounds, Wilder thought. It was about the size of a wildebeest. And related to other hooved mammals like the horse and rhinoceros. It maintained a diet of herb and roots.

The animal inched its way past Wilder. Then it paused for a few moments. When it became aware of his presence, it skulked off into the jungle like an out of control locomotive.

Wilder continued pressing on.

But hunger and thirst were beginning to be unbearable. Then he remembered his days with Interpol in Rwanda. He was not particularly afraid of the two men hunting him down after they killed a local agent and thought they had wounded Wilder. But he hid out in the mosquito-infested underbush and dark foliage in that African rainforest for days. His trackers didn't know where he was. They called off their hunt and vanished.

Though he most likely recalled how he stayed alive. It was the hog plum that he nibbled on to allay his pangs of hunger. The fruit grew wild everywhere. Particularly along river banks, tangled masses of vegetation, tropical scrub and thorn bush forest.

Wilder charged ahead until he found the hog plum. He sank his teeth into the fruit. Its heavy acidity caused a sudden spasm to ripple throughout his body. The taste was strange. Just as he remembered in Rwanda.

Minutes later, he found the water vine. It embraced the trunks of cedar, purple heart, balata and large trees. Rolling on and on towards their boughs like a dense, leafy ligature.

Wilder quickly cut one of the vines in two places, to prevent the water from being absorbed back into its vital center. Then he ingested large amounts of sweet, clear and delicious water.

Then his cell phone rang.

The unexpected sound made him blink with surprise.

"Hello," he quickly answered.

"Hi…Frank," was the reply.

Wilder knew it was Hyman's voice. It was constricted and full of tension.

"Phillip! What's goin' on? Where are you guys?" Wilder asked reeling with amazement.

"Man, gimme that phone!" It was the sound of a strange exploding voice from Hyman's phone.

Wilder was quiet as a tremor of apprehension flashed across his body.

"Are you Wilder?" a voice blurted out with brutal detachment.

Wilder thought for a moment. He never heard the voice before. Had something gone wrong with Hyman and Kumar. He wondered.

"Yes. I'm Wilder."

"Ok…Wilder. You listen to me. And listen good. My boys and I got your buddies here. Hog tied. We sneaked up on them last night. One after the other. If you got any sense, you'll be here as fast as your feet can move through this jungle and give up. You won't get hurt."

"Sure, sure thing, man. Who're you?"

"You really wanna know, constable!"

"Yes. Won't hurt to know the man I'm coming to see."

"Name's Maxwell Barnes."

Then Wilder remembered seeing the man's mug shot and wanted posters around Georgetown. He was about forty-six and one of three surviving members of a group of five prisoners who stabbed and shot their way out of the Camp Street jail six months ago. They ushered in a campaign of terror across the country. In four months about nine policemen as well as civilians were killed because of the rampage orchestrated by the escaped felons.

Two members of the group were gunned down in a sugar cane field at Providence on the East Bank of Demerara.

And for six weeks nothing much was heard about Maxwell Barnes and the other two men. Rumors circulated that they were living with friends somewhere in Wismar, down the Demerara River. Then Ituni and Sophia.

"Ok, Maxwell. You don't have to hurt them. Just gimme your location and I'll be there."

"That's what I wanna hear, Wilder. No one will get hurt like in Georgetown and the East Coast. We're in this for some gold and some money. And we're outa here."

"Sound like you'll be a man of your word, Maxwell. I don't want trouble myself."

"Look, Wilder, there's nobody in this whole damn jungle who will give Maxwell Barnes any fucking trouble. See this AK-47 I took from your buddy. Man, it can take out the whole damn police force in Georgetown."

"You aren't kiddin' man. Where are you located?"

"About four miles east of Tumatumari Rapids."

"I'll be there."

"Now lemme say this, man. Don't know you. Never met you. But whatever you do, don't try no shit. Else you all will be three dead constables!"

"I mean what I say."

"Ok fella. If you want to save these two constable's lives and yours, you better get your ass here before it gets dark."

"You're the man, Maxwell."

Chapter 14

In Wilder's mind, fear could be a powerful motivator. After years of service as a Newark cop and an Interpol agent worldwide, he understood the reality of fear. And the havoc it could wreak if left unchecked. Fear begot panic. It could cause a man to act in ways inconsistent with his judgment or character.

As a rookie cop, he recalled the day in Newark, New Jersey when he was confronted by a man wielding a machete. Fear invaded his mind. His lungs got tight as if all the air was knocked out of his body. And suddenly it was as though he couldn't breathe. But quickly he was able to dispel his tension and handle that adversity based on his indoctrination as a cop.

That epiphany taught him how to harness his emotions under the most excruciating circumstances. And to make viable split-second decisions when events arose.

And there was Maxwell Barnes. A former porknocker turned felon. A killer of police and an individual chronically involved in banditry. A man who anticipated that Wilder would forfeit all his police expertise by signaling that those acquired knowledge were negotiable in a situation of grave and imminent danger. Wilder knew that Barnes was far from being sincere. Why should he like a blinded idiot take the word of a madman, a piece of crap and endanger the lives of his comrades and himself.

That day in the Essequibo, Wilder didn't have to reach too deep into himself to figure out who he was and what was going to be his response to Barnes and his cronies.

He checked his weapon.

It was a 9 mm Browning high-powered pistol. A primary weapon for all security officers at Omai gold mine. Created in a different era, sometime in the late nineteenth century in the state of Utah, the 9 mm was considered by some a relic. But after years and years of constant makeovers, the firearm had achieved a state of the art metamorphosis. The end product was appropriate for Wilder's discriminatory standards.

To him it continued to do what a good weapon did.

It never failed him in Africa or the Caribbean. Because it put the holes where he wanted them to go.

Wilder continued south towards the sound of rushing water. He looked backward and to his left and right. Straining his eyes to catch any sudden movement among the bushes. Half hour later, the rushing water sound grew louder. As he closed in, the sound intensified. Wilder edged his way through masses of thorny undergrowth, reeds, and grasses. They made travel difficult. But he pressed on.

Panic nearly consumed his mind as he reflected on his situation. Thousands of square miles of dense forbidding jungle where his life was hung in a balance. And he reckoned that to tread in the jungles of Guyana alone, was to know fear. He continued his prowl.

Then he encountered a line of trees and rocks. He had to slow. Carefully placing his feet in the right spot, for fear of injuring a leg.

Then the dark coffee-colored water of a creek emerged. Wilder kept out of sight as he scrambled and stumbled from rock to rock, headed in the direction of the rapids.

He heard the rushing waters. And crouched between some tufted plants where thorns predominate. No signs of life between distant trees. He slipped into bushes and trees and continued creeping towards the rapids. For several minutes he lay amongst the thornbush with a good view of the Tumatumari rapids.

There he lay eying the alternate masses of grass and bush along the opposite bank of the creek.

Suddenly in the distance, he knew he saw a figure. It was a man, Wilder figured. Though hard to distinguish.

Then he saw it was two men. Sitting by the creek and smoking cigarettes. Wilder figured that the third man might have been guarding Hyman and Kumar. Though it wasn't their number that bothered him. What did, was that they were in possession of some heavy fire power after relieving the constables of their weapons. An AK-47, a Mossberg shotgun and two Browning 9 mms.

Wilder had to cross the creek in order to engage the men. They were on the other bank. He lay on his stomach formulating a plan.

He wiggled back about three hundred yards from where he had come. Gliding his way through shrubs and tall grass. Soon he was where he wanted to be. An area of the creek where the banks had narrowed.

Might have been about 3 p.m. Already mosquitoes and gnats were raising hell. A smoky, cobweb-white of a mist was already cuddling the tree tops. Daylight was on a frantic retreat.

At that point of the creek, far from the rapids, the velocity of the water flow had decreased. Almost to some degree of stagnancy. Wilder knew that such an environment was fitting for piranhas. Those deep-bodied carnivorous fish. Noted for their voracity. Some grew in lengths of up to two feet. But all had sharp saw-edged teeth that closed in a scissor-like bite.

Wilder figured that getting to the other side of the creek wasn't just a walk in the park. He had to make sure he didn't attract the attention of Barnes and his bunch.

And the killer fish. Another personification of danger that lurked. A predator that attacked with speed and surprise. With an advanced sense of smell and vibration in water that would almost certainly sever Wilder's lower extremities the minute he laid foot in the dark brown liquid.

He arched his body to a crouch and gazed towards the rapids. He was unable to see the men. They were hidden from view by the outcrop root of a giant greenheart tree.

His thoughts were in high gear. He had to create a distraction for the piranhas and cross the water fast. But he didn't want to alert the bad guys of his presence.

Then the idea came to him. What if he threw a piece of rock in the water? The piranhas would, by instinct, rush towards the rock in droves. And it wouldn't take them long to realize that the rock was only a decoy. That meant he wouldn't have enough time to scamper across the creek.

Wilder knew that a piece of meat was the only lure to keep the piranhas away from him.

Then his cell phone rang.

"Hey, constable! Where are you now?" the bellowing voice of Maxwell Barnes echoed in his ear.

"On my way. Gimme some more time," Wilder said, making it sound apologetic.

"Hey cop. You better hurry. You don't have much time," Barnes responded with a voice cracking with fury.

Wilder knew that Barnes and his cronies wanted Hyman, Kumar and himself dead. He was morbidly aware of their intentions. By taking everyone out, the trio would make good their escape. And no one would be left alive to bring them to justice. They would vanish in thin air.

Wilder's main concern at that time was to cross the creek without being eaten by piranhas. He had some hunting skills. And a firm notion of what he was going to do, crept up in his mind. He raced from the creek and into the jungle. There the mosquitoes were relentless.

He stood at the edge of a wooded area for a few moments. Then among some bullet-wood trees overhead, he heard a sound. Wilder gazed overhead and at first thought that the animal that was moving higher towards some branches was a marmoset. But careful examination made him realize it was a sloth. As the name implied, it was a slow-moving nocturnal mammal about the size of a cat with thick brown fur, a short flat head and big eyes. Its claws are sharp and menacing and the only defense against predators like the jaguar and the harpy eagle.

Wilder scuttled up the bullet-wood tree with almost a feline grace. He succeeded in breaking the branch of the tree where the sloth hung with some difficulty. The animal's ferocious claws clutched the severed bough in a death-like grip.

Wilder ran towards the creek in a crouch as the animal struggled to maintain its balance on the branch. It gazed at Wilder with taciturn eyes.

Then he got to the edge of the creek. His heart beating wildly. Spurts of adrenaline coursed through his vein. Thoughts of wading across a piranha infested creek caused his knees to quake.

But, then he had no time for second thoughts. It had to be done.

One. Two. Three.

Swoop!

Wilder hurled the branch with the sloth almost to the center of the creek. Momentarily he dashed towards the water. Waist deep. Bounding, charging towards the opposite bank. His weapon and his cell phone held high. Never looking back. All he saw was a feeding frenzy with his peripheral vision. The water was very cold. It felt as though there were pieces of ice in it that had just melted.

And it was brownish, like coffee with no milk. Smelling fresh and disintegrating like glass around him.

Wilder powered his way towards the bank. It was a hard thirty-five-feet dash against the subtle current and near-rapid insistence of the water.

He turned his gaze to the center of the creek. And the sloth. Its thick brown coat was no match for the scissor-like mauling of the piranhas. Scores and scores of the carnivorous fish hacked and lacerated the helpless animal as they rose to the surface in their violent feeding frenzy.

The area of the encounter grew brownish red with the sloth's blood. And within minutes, its skin moved randomly above the water. Bobbing to the left and right as shoals of red-bellied piranhas mangled the pelt for bits and scraps of morsel.

Then the ordeal ended. For only the first in a bundle of obstacles, he thought. The solutions were getting difficult and more difficult.

He scrambled and shifted around, getting the water from his boot and wringing his clothes almost dry to eliminate the physical discomfort of them sticking to his body as he moved.

Then he became aware of the sound of his breath as he ran and walked towards the direction of Barnes and his men. And his captured comrades.

Chapter 15

The prospect of a firefight was not the first of Frank Wilder's career. No. Not even reckoning that was long sought. He had never wondered what the next step was, during an encounter. His reaction was automatic. But that day, deep in his psychological subterrain, he was gripped by an uncanny feeling of ambivalence as the anticipation of a gunfight loomed before his consciousness.

Why did a young man, just after college, choose a dangerous career as a member of the Newark Police Department in New Jersey, then join Interpol and finally the Guyana Police Force? Was it to stand one day face to face with his own mortality, to suffer the ghastly gaze of Death, to understand that its gloomy eyes were ultimately his own? Might that have been his reason to come back to Guyana and die in that stinking green hell of a jungle? He had no answers.

But quickly he dismissed his musings when he realized that as a mortal, he had to revere tracking a trio of killers. And as a brand new constable, it was a brutal rite of passage, the culminating test of his monomaniacal, obsessive training. As a human being, he undertook his task as a spiritual trial. Regardless, it had to be done.

It was four in the afternoon. Wilder forged ahead. He stood at a part of the jungle where he saw the sky. There was no sunlight. The sky had grown dark. Suddenly a renegade wind as if from nowhere battered the tree tops. And the heavens opened, scattering bolts of lightning and hail across the land. As if a giant prehistoric beast was moving across the jungle.

Far in the distance, monkeys howled. Jaguars roared. And wild hogs squealed. Wilder was almost hit by a falling branch as nature wreaked havoc in the jungle. Wilder groped by the trunk of a mahogany tree, as iguanas, snakes and the young of birds fell before his eyes.

But he was desperate to continue his journey. Should he venture out, he thought. But he realized a blow from a broken branch could lay him cold. But the longer he stayed hidden in the bosom of the mahogany, the lesser would be his opportunity to sneak up on his adversaries.

Twelve minutes elapsed.

The tropical storm continued its ferocious romp throughout the landscape.

Wilder waited, looking to his left and right with a feral glitter in his eyes. As if expecting someone. Something. Anything.

Twenty minutes later the wind abated. And the storm died.

Wilder worked his way towards the Tumatumari Rapids. The absence of the wind and the rain afforded him renewed energy. But along the way, his body sagged. He felt as if his feet were dragging.

Then he heard voices.

He crawled on his belly towards the sound.

What he saw next caused his stomach to contract to a tight ball. The men were sitting on tree stumps and logs in a cleared away area. Three of them. Wilder figured it was Maxwell and his men. Kumar was lying on the jungle floor, on his back as if he was clutching to life. While Hyman was tied against a tree. His head hanging down. Blood was on his shirt. Wilder figured they might have been tortured by the bandits.

Four bags of reclaimed carbon were sitting in the area. One man smoking a cigarette had Kumar's Mossberg shotgun and 9 mm by his side. Wilder reckoned it might have been Maxwell. He did all the talking. Besides, he sounded bossy.

Another man had an AK-47 that belonged to the sergeant, while the other crook had his 9 mm Browning.

Other items like bed rolls, rations, a pick axe, a shovel and two machetes littered the area. And the place was almost flooded by rain water from the storm.

Wilder checked his pistol to make sure it was there. Then a spasm of indignation flashed across his face. He knew what had to be done at that moment or never. The thought of it made his blood surge wildly through his body. But he knew there was no other way.

A quick plan ran across his brain. He had to take out the man with the AK-47 first. He couldn't afford to be shot at with such fire power. There was no cover in the jungle that could save his life. He once saw the AK-47 penetrated the engine block of an SUV in the streets of Newark. And killed a man. Those bullets would bust open the trunks of trees. Again he interrupted his thoughts. Why did he have to ponder it so long. If that continued he might just want to chicken out. Hell no, said his little voice inside. Those bastards would not get away with what they did to his buddies.

Then Wilder reckoned he had hesitated too long as it was. Quietly he inched his way closer to the perimeter of the camp. Against a fallen wallaba tree, where mud and rotted leaves, mixed with earthworms pressed against his chest. He wiggled forward and sideways another couple of inches. And hesitated just to look and listen.

Maxwell was speaking as he puffed away at a cigarette. His whiskey-roughened voice was throaty. As if he was aware of Wilder's presence somewhere. And didn't want to give away his position. Maxwell told his men that once the other cop showed up, all three would be slaughtered. He said that he would take out Wilder while his men would finish off Hyman and Kumar.

Wilder remained still. Fighting his inner panic to get the situation over and done. He concentrated on the men in the hollowed out area.

The raw venom in their eyes seemed to ignite Wilder in mind and body, like the thermal effect of a forest fire. He lowered his glance, avoiding an inadvertent eye contact with the gang. And suddenly he became aware that his body was about to quiver.

Bang! Bang! Bang! Bang!

His Browning 9 mm barked.

Wilder's bullets went just where he wanted them to go. One shattered the long bone of Maxwell's upper right arm. Spilling his blood, while the wounded arm dangled into paralysis.

Another bullet ripped through another man's shoulder blade. The impact sent him reeling backwards in pain.

Wilder disabled the AK-47 as one of his bullets knocked out the trigger. One man made a dash for a fallen 9 mm. But he was too late. One of Wilder's bullets lifted the weapon from the ground and sent it ricocheting into the jungle.

Maxwell and the other wounded man howled with pain as they lay cowering in the jungle.

Wilder sprung from his cover after getting off four shots.

The third man, unwounded, made a frantic dash for the Mossberg shotgun, not far from his injured leader.

"Go ahead. You son-of-a-bitch. I'll bust a cap up your ass if you touch that shotgun!" Wilder's voice boomed out, growing thick with insinuation.

The man quickly abandoned his quest to gain the shotgun. He looked at Wilder with smoldering eyes. His face was fleshy with a square jaw. And his neck was as tough as beef jerky.

The man eyed the 9 mm pointed at him with a withering stare.

"This is the end of the road, pal," Wilder said, his voice less intense.

"That's because you've got a weapon," the man responded with a husky voice.

"I'll put away my gun. You wanna take me on."

"Damn right man! I'll beat your butt silly. Because you're a cop, you think you'll push people around, eh. Won't work with me, shit head!"

"Ok," Wilder said, putting the 9 mm between his belt and abdomen. "Make your move, wise guy."

The man surprised Wilder when he opened his shirt, exposing the handle of a twenty-two inch cutlass. He withdrew the weapon from his belt and circled Wilder as he muttered peevishly under his breath. Then he moved to Wilder's right, then to his left.

Wilder fixed him with a level stare.

The man again circled to Wilder's right.

Wilder's peripheral vision raked the hollow. His feet nimble, ready to move like a beast of prey.

Suddenly the man's hand exploded in a blinding flash as he dashed towards Wilder. Ranting and raving. The cutlass whistled and sliced the air with malicious intent. First to Wilder's body. But instinctively, he drew aside. Then to his feet. Wilder went airbound like a balloon in levitation. Avoiding the blow.

But the man continued his advance in a one dimensional attack, shouting obscenities at Wilder.

Wilder studied his method. He timed the man after he exerted all his energy, missing blow after blow. Then the deadly cutlass severed the trunk of a wild pine tree. The man's body swung to its right, almost to an angle of one hundred and eighty degrees. Exposing his right rib cage.

Wilder capitalized on the man's split second mistake. Quickly he became the aggressor with that small window of opportunity. It was a brutal side kick that Wilder unleashed. With stunning accuracy. The blow caught the man on his right abdomen. Just below his rib cage. The man winced with pain as his body recoiled.

But immediately the man recovered, with the quick resiliency of a wounded animal.

"I'll fuck you up, Wilder!" the man shouted, his voice growing maniacal with animosity.

"Give up man. Don't let me hurt you," was Wilder's response.

"No. Never. You cocksucker!"

Wilder blinked with surprise as the man sprang towards him, despite his injury. The blade of the cutlass missed Wilder's chest by centimeters. It severed his shirt with a clean, horizontal gash.

Wilder engaged the man before he unwinded. Quickly he grabbed the man's wrist with his left hand, in a frenzied attempt to relieve the man of his weapon.

But the man mounted a conscientious effort to hold on to the cutlass.

At that time all hands were locked on the man's hand with the weapon.

They struggled in the hollow. Knocking over bed rolls, hammocks, and boxes of ration. Then they moved back and forth to buttress roots of mahogany, pine and purple heart trees, and among smaller trees and shrubs. Wilder kept his grip on the man's cutlass hand and succeeded in bringing his own left palm to the rear of the man's elbow. He continued his vise-like grip on the man's wrist with his right hand.

Then the man stiffened up his lumpy body. And at the same time, pushing his massive oarsman's shoulder against Wilder's breastplate. Wilder succeeded in avoiding the man's loose left hand as his calloused fingers ripped away his shirt. His high-domed head with wooly black hair jabbed at Wilder's jaw.

That annoyed Wilder.

But what was more annoying was the smell of the man's dragon-like breath. It was hot in Wilder's face. He figured the man was faced with some serious issues.

Wilder mustered all his energy. He had to end the altercation quick. He pushed the man's elbow forward and simultaneously pulled back on the man's wrist.

Wilder's opponent gritted his teeth with pain.

"Holy mackerel!" the man hollered. "You're breaking my arm, you dog!"

Wilder did not respond

He intensified his effort.

And crack!

There was the sound of breaking bones.

The man's elbow was snapped.

The cutlass instinctively fell from his grasp. He grimaced in agony. And bawled. And squirmed.

Wilder kicked him up the romp as a parting blow. It sent his adversary reeling across the hollow and flat out among ferns, mosses and herbaceous plants.

Tears welled up his eyes as he moved, locking his body into a fetal position.

"Shit! You broke my hand. You jackass!"

The man grumbled his words with difficulty.

"You got off light, mister. I should have broken your neck!" Wilder answered as a scornful grin ran across his face.

Chapter 16

Harold Radford became chairman and chief executive officer of the Wadpole Corporation sometime in 1992. His goals for the mining giant, the second largest in South America, had remained very simple. Strategic growth through expansion in the rain forests. And an image makeover that stressed Wadpole's diverse capabilities.

Radford had been chief operating officer of Toronto Chemicals. A Canadian company. He spent twenty-two years with Edmonton Shipbuilders before being tapped to lead Wadpole, a Montreal based company with subsidiaries world-wide including Guyana, Peru and Costa Rica.

Some described the fifty-eight year old Radford as being methodical and focused. They said he helped Wadpole make a smart move, when the company invested a half of a billion dollars in the Omai venture. A venture that earned Wadpole six-hundred and sixty-one dollars for an ounce of gold at the London and Zurich stock markets and the New York Mercantile Exchange.

Radford's methodical business style explained why he wrote a memo to the Human Resources Manager in Brickdam, requesting the transfer of a rank from the Omai gold mine security force to Wadpole's Headquarters in Georgetown.

That rank was Frank Wilder.

It started two months ago, when Wilder single-handedly took on a trio of criminal porknockers turned thieves who had stolen reclaimed carbon from the company's mine site in Omai.

According to the Georgetown Evening News the crooked porknockers ambushed a veteran police sergeant and corporal. Took their weapons, tied them to trees and tortured them.

NCN Channel 11 Demerara and DTV Channel 8-Berbice had nothing but praises for Constable Wilder. They described him as a Guyana 'Rambo' who infiltrated the criminals' hideout in the Essequibo jungle with stealth, speed and agility. Then rescued his comrades.

The news described how he severely injured Maxwell Barnes, a notorious cop killer and reputed leader of the gang. And also crippled his two cohorts although they were armed with two 9 mms, an AK-47 and a Mossberg shotgun, taken from his comrades.

Wilder had become an overnight sensation after he loaded the blood-soaked crooks and his delirious buddies into a boat belonging to the bad guys and headed across the Tumatumari Rapids towards Mahdia. It took two days for soldiers from the Guyana Defense Force to locate him.

From Mahdia, the men, including Wilder, were air lifted to Ogle then to the Georgetown Hospital by ambulance for emergency medical treatment.

Wilder was discharged from the hospital in two days and resumed his duty at the Omai gold mine. Maxwell Barnes and his lot were remanded to the Camp Street jail after a week in hospital and were awaiting trial. While Hyman and Kumar, after a brief hospital stay, were placed on sick leave.

The following week, Wilder got a visit from Harold Radford and Roger Balkarran, Omai's security chief.

The man selected to head Wadpole met ranks at the Omai conference room. He voiced an optimistic proclamation that Omai's time for turning up the heat on pesty thieves had come. He praised all ranks for their hard work and diligence.

"We thank heavens that Constable Wilder was there at the Tumatumari Rapids at the right time," Radford said as his face beamed with good cheer.

Then Wilder received a long standing ovation.

Months went by.

Wilder continued his duties at Wadpole's head office in Georgetown, watching for encroachers, deadbeats and criminal elements who might venture on the company's property. Especially when shipments of gold were coming from Omai.

But that was only a part of Wilder's problems. He also was wary of internal pilfering of gold-bearing material by mine and office workers.

One day he arrested a yard employee for indulging in petty theft from the company's warehouse on Alexander Street. The stupid goof asked his buddy to park a truck on the Charlotte Street side of the company's property, anticipating that Wilder wasn't going to make his rounds for fifteen minutes. But Wilder smelled a rat and made a bee-line along his path. The man, from South Ruimveldt and an employee of Wadpole for the past three years, was caught red-handed. He was seen passing such items as tires, cutlasses, and diverse amounts of construction materials to another man through the fence. The man on the other side of the fence was loading the stolen items into a Toyota pick-up. Ready for the black market. But Wilder foiled their plan. Eventually both men were cooling their heels in an eight-by-eight jail cell on Camp Street.

The following day news came to Georgetown that Dr. Sugrim Bhoopat, deputy to the Minister of Health, was killed by a cutlass wielding Amerindian in Micobie Village, far in the Potaro region.

Investigators, according to the police, revealed that Bhoopat had gone to the area in response to a growing number of residential complaints about flooding conditions and the possible outbreak of leptospirosis.

Sources said that the chief physician at Mahdia Regional Hospital had cautioned his staff to stay on high alert after reports of suspected leptospirosis cases increased fourfold in just a couple of days.

The hospital sent a fax to the Ministry of Health in Brickdam notifying the agency that there had been a rise in rodent infestation in residential communities after heavy thunderstorms flooded several villages about two weeks ago.

Time passed.

No one at the Ministry responded to the faxed message

Meanwhile, in Micobie Village, outside of Mahdia, residents complained that there was an increase in fly and mosquito infestation since the thunderstorms.

The head of Mahdia Hospital said that his facility was unable to handle the volume of patients with symptoms of leptospirosis. People, he said, were waiting too long before seeking medical attention. The physician confirmed that his facility admitted twenty-five patients. But six died. He pointed out that the last patient who died had sought medical attention within minutes of his demise.

That was the incident that prompted Guyana's Minister of Health, Doctor Dennis Harrypaul, to send his deputy, Sugrim Bhoopat, on a fact finding mission.

But that fact finding mission caused Sugrim Bhoopat his life.

Eyewitnesses said that Bhoopat had gone to the town's hall that Friday for an impromptu meeting with residents of Micobie Village. He was accompanied by three armed policemen and two of his aides.

Sometime after 2:30 p.m., during a question and answer session, an Amerindian man about forty-six years old, became enraged. He looked at Bhoopat piercingly and accused the official of causing the cyanide spill at Omai in 1995. He said the disaster killed his family and his cattle and drove him out of Omai. He continued to yell at Bhoopat with a voice high and hysterical.

"You killed them all, you dog. Now you're trying to kill me."

Just then, the armed cops closed in on the angry man. They beseeched him to be calm. And to keep his voice down.

Bhoopat's face showed a glare of surprise as he swallowed hard on his saliva.

"Let's talk about your problem, sir. I can understand your anger," Bhoopat said, appearing calm and trying to conceal the chill running up his spine.

But the Amerindian fanned the air with his hands in a show of disgust. His face bearing a stony expression. And without warning, he unleashed his feral instincts.

Quickly he drew a cutlass that was cleverly tucked under his clothes and sprang at the deputy minister of health with his face bearing a predatory expression.

The weapon sliced through Bhoopat's right jugular. The official attempted to block the second blow as he reeled backwards. But without success. The cutlass hacked off his hand.

Everyone was caught by surprise as the attacker catapulted through a door and onto the street. A crazed, insipid expression lingered on his face.

Cutlass in hand, he attempted to flee the scene. But he was cut down in a hail of bullets fired by the cops, in their attempt to bring him down.

Sugrim Bhoopat died on his way to Mahdia Regional Hospital.

When the news reached Georgetown, Dennis Harrypaul, the Minister of Health, gawked in disbelief.

Appearing to be livid with rage, he balled his fingers into tight fists. And hammered them into his desk.

"How did this happen? Where were the friggin' cops!" he screamed with a voice growing hoarse and husky.

"People said the Amerindian had a cutlass hidden under his clothes, sir. No one expected him to attack Mr. Bhoopat," an aide answered in a constricted voice.

"I don't give a rat's ass. Those damn policemen had guns. They should have been able to stop that man before he got to Sugrim!"

"What are you planning to do, sir?"

"You'll see. Get the Ministry of Home Affairs on the phone now. I want those constables fired for dereliction of duty."

"Sure, sir. But we've gotta replace them as quickly as possible."

"Guess you're right. I do need a good cop to maintain my safety. Especially when something like this happens."

"Anyone in mind, sir?"

"Yes. Remember that new cop who beat up on those porknockers at Tumatumari Rapids."

"Yes, sir."

"His name's Wilder. Frank Wilder. We need that guy right now. Do you hear me?"

"Sure, sir. Sure."

Chapter 17

If there was a place in Georgetown set aside for the rich and powerful, Frank Wilder moved into it. That day he hauled his suitcase and everything he owned. The place in Prashad Nagar was quite a departure from some of the places he was forced to call home. Unlike the Margarita Hotel, the police dormitories at Eve Leary and Omai and the flats he occupied at Potaro Landing, Prashad Nagar was better living than those places, Wilder figured. And by a heck of a lot.

After the unprovoked attack on Doctor Sugrim Bhoopat that left him dead, Dennis Harrypaul, the Minister of Health, was running scared. Sources said that he had received death threats from splinter factions of the Guyana Organization of Indigenous Peoples. But leaders of GOIP disputed that claim.

It was at that time that a call was made to Harold Radford's office in Robb Street. The caller was Roger Balkarran, the man in charge of security at Omai.

Radford picked up the phone.

"Mr. Radford. This is Roger Balkarran from Omai," his crude voice sounded over the wire.

"Hey, Roger. What's the occasion?"

"Regret to say, Mr. Radford. Just got orders from the top. They will be taking Constable Wilder from you."

"Why? For what reason?"

"You and I both know that Dennis Harrypaul was pissed after his deputy was killed two weeks ago in Micobie Village."

"Who wouldn't be, Roger?"

"Well, he blamed the three ranks for not reacting in time to save Bhoopat's life."

"So."

"The three cops got fired."

"Ok, I see the picture."

"You're right. They're replacing these cops at the Ministry of Health. And Wilder is one of them."

"Look, I hate to see the man go. But there isn't a damn thing I can do. He is a good policeman," Radford confessed.

"You're not joking. Anyway, the rank who will be replacing him at your office will be there tomorrow."

"Ok, Roger."

The next day Frank Wilder was transported by Land Rover from Wadpole's head office on Robb Street to the Ministry of Health situated on Brickdam Drive, in Georgetown.

A police sergeant directed him to the office of Doctor Dennis Harrypaul, the Minister of Health.

He walked towards a plate glass door and opened it. An Indo-Guyanese woman sat at a desk. Wilder figured she might have been Harrypaul's secretary. An attractive woman, Wilder thought. Shoulder length black hair tinted with a touch of scarlet. And smoothly coiffed to curve over lightly brown cheeks. Her unshaded eyes appeared moist and luminous and her mouth made tantalizing by an oval accent of the full lower lip.

"Hello," she said, her face beaming. "Mr. Harrypaul has been expecting you."

"Thank you," Wilder answered.

Then the woman rose from her desk. She walked towards the door of the Minister of Health, aware of her erectness. And conscious of the limber movement of her shapely legs and the upturned firmness of her breasts.

Within a few moments she was back.

"Mr. Wilder," she said. "The Minister is ready to see you."

"Ok."

Wilder opened the door and walked resolutely across the carpeted floor. Dennis Harrypaul was an aging man about sixty-five. His dark, tired eyes appeared to have been wedged open by a jackhammer. Scanty clumps of graying hair were carelessly plastered to his scalp. And his ears, far out of proportion with his body, were glued close to his head. His hands were large and meaty, with thick, calloused fingers. Uncharacteristic for a man of his office, but with incredulously manicured nails. He wore a shirt of blended silk. Short sleeved with unmatching tie.

"Thanks for being here, Constable Wilder."

"It's a pleasure, sir."

"Heard you're very good at what you do."

"Always try to do my best, Mr. Harrypaul."

"I know you're aware of the incident at Micobie Village."

"Yes, sir."

"Now, you and I both know that my aide, Mr. Bhoopat, didn't deserve to die that way."

"What you mean, sir?"

"I mean that if those ranks were doing their job, my deputy would have been alive today. That's no lie."

"I see your point, sir."

"Constable Wilder, all over this country, Amerindians are bitter and resentful of this government and the gold mining company at Omai because of the cyanide spill. And because some tribes had to be forcibly removed from their land by eminent domain to benefit the mining and logging industries. During this whole ordeal, many people lost their lives. That was unfortunate. But Guyana must move on. Our economy thrives on these ventures. And in this climate, we as government officials cannot let our guards down."

"Where do I come in, Mr. Harrypaul?"

"According to your file I have before me, you're a damn good cop."

"Thank you, sir."

"That's why I notified the Ministry of Home Affairs to have you transferred here as my guard to protect me here and during my contact with the public."

"That's ok with me, sir. But do you think it is really necessary?"

"You bet it's necessary, Constable Wilder. I have very many enemies, you know."

"No, sir. I don't know. But I guess you've found the right man for the job.

"That's what I wanna hear."

Chapter 18

Evening twilight came to Georgetown.

It was three weeks since Frank Wilder was reassigned from Wadpole's Headquarters to the Ministry of Health.

That day he left the Ministry close to 7 p.m. He got into the Toyota Land Cruiser. The big three-fifty-engine roared when he cranked the ignition. From Brickdam Drive, he continued east towards Vlissengen Road. He was headed for his apartment in Prashad Nagar. Wilder felt he wasn't doing too bad at that point as a feeling of euphoria ran through his mind. First he was given a one bedroom flat in a classy section of town. And before he knew it, someone threw him the key to a Toyota SUV. On top of it all, Wilder did not spend a penny. The taxpayers were picking up the tab.

Then he got to thinking of the man who was behind the whole scenario.

Doctor Dennis Harrypaul.

A man who was considered the chief medical officer throughout the entire country. An official, Wilder surmised, who had suddenly acquired a morbid fear of the public, after the cyanide spill at Omai. And more so, after the violent death of his aide, Sugrim Bhoopat.

But Wilder could not see the link. Cops were all around the ministry. The issue puzzled him, catapulting in his mind. Over and over again.

Wilder sensed that something clandestine was in the making. Or might have already been made. He came to that conclusion after he last saw the man, that noon time after he, Harrypaul, exited his office briefly and returned.

His face bore a dubious expression. Like the proverbial cat that ate the canary. As if he wanted to free himself of his torment. A torment that was buried deep into the façade of a seemingly shrewd, unfeeling man with a possible flair for the occult.

Those were only Wilder's musings.

Then he reflected on Sugrim Bhoopat. The poor bastard apparently didn't stand a chance. And Wilder knew long ago that dead men told no tales. Because death was too unconditional to be ambiguous.

But Wilder guessed that someone had to make a start with a dead man.

No. He didn't mean the sleuths at police headquarters on Brickdam. Men sitting on their obese backsides, drinking coffee and tea and smoking fat cigars.

Hell no.

That would be a conflict of interest from the beginning. For those guys, the circumstance surrounding the death of Sugrim Bhoopat was already investigated. And determined to be a senseless murder committed by an insane, heinous man.

For Guyana, that case was closed.

Besides it was taboo for one ministry to investigate another ministry even if there was strong evidence in any situation of foul play or impropriety.

It would be swept under the rug.

A strong wall of silence existed amongst government ministers. Not even President Charles Jagnarine questioned any of his underlings. Except if someone was pointing a finger at the chief executive himself.

And that was where Frank Wilder, a former city cop from the USA, turned undercover as an Interpol agent to investigate the facts of the Omai cyanide disaster unknowingly came in. To secretly investigate Bhoopat's death and determine its connection to Dennis Harrypaul and the handling of the cyanide spill in Omai. His search might lead him to bigger fish. Like the president himself. No one knew for sure.

It was 7:30 p.m. that evening.

Night had already fallen on the city. Wilder figured he would make a stop at one of the clubs in Subryanville and get a beer before heading to his place.

He made a left on Sheriff Street. The traffic was thick. And before long he found himself stuck in a convoy.

The rush hour was far from over. Sheriff Street was a composite of horns and engines and voices shouting at other voices. Sputtering obscenities.

The lingering smell of exhaust fumes and unburned gas and diesel fuel hung in the sultry air. Traffic lights that were designed to keep traffic at a normal ebb and flow seemed to be ineffective. And always there was a minibus, truck or car stuck ahead. People, too, at the crosswalks would take their own sweet time to navigate Sheriff Street. While others hauled pushcarts, baby carriers or broken down mopeds in their frantic scurry to clear the street.

Ordinarily, from the Ministry of Health to Subryanville, traveling would not take more than half hour. But not that night. It was 8:25 when Wilder reached his destination. He pulled the Toyota to a sudden stop and parked in front of a place called Buddy's Nite Spot.

The heavy beat of reggae music grew louder as he neared the building. And a couple of hookers, street crawlers and drug dealers littered the sidewalk.

Wilder meandered around them.

No breeze was blowing from the Atlantic. The air was hot and sticky. And the humidity was punishing.

He entered the club.

The dramatic change in atmospheres was like day and night. The place was cool. High ceilinged, punctuated by the softness of rainbow-colored light. But the floor was filled with wall to wall dancers. They appeared oddly frozen under the flashes of small multicolored lights.

Wilder pushed his way to the bar and ordered a drink. The man behind the counter appeared to be a silent old dog. Without saying a word, he pushed the drink in front of Wilder. Whisky and ginger. Wilder picked up the drink and sipped it. His sense of smell quickly picked up the pungent-like odor of marijuana. Despite looking around him, it was difficult for him to figure out who was smoking the weed. Dancers and party animals were everywhere. Smoke filled the air like a backyard barbecue.

Then he heard a voice behind him.

"Do I know you?"

Wilder spun around. There was a woman standing there. Young, about thirty-three, with shoulder length black hair. Eyes moist and luminous and a full lower lip.

"Hey, Alma. What wind blew you here?"

"My girlfriend and I sometimes stop here for a drink and listen to music."

Alma Rampersaud was the executive secretary to Dennis Harrypaul, the Minister of Health. She had been working with the ministry for the past three years.

"Where's your friend?"

"She stepped in the ladies room."

"Ok, the drink's on me."

"Thanks. I'll have a vodka martini."

"You got it."

"Gee, Frank. It's quite a surprise. Never thought I'll run into you at Buddy's."

"Tryin' to get familiar with the neighborhood."

"That's right. You left Guyana at twelve. So you said."

"Yes. I was very young."

"Where's your family? You never mentioned anyone when we spoke in the office."

"Both parents are dead. I've got one sister. She's married and lives in Brooklyn, New York."

"Oh. I'm so sorry to hear about your parents. Heard from your sister lately?"

"Not recently. Gotta call her. How about you? Got a boyfriend?"

"No. Was once in a serious relationship, but things went sour. We both went our separate ways."

"What about Dennis? Notice he eyes you like a fox by the henhouse."

"What a jealous old fart. He's married but gets an attitude if someone tries to make a pass at me."

"I picked up on that long ago. That's the reason I stay at my desk most of the times and avoiding making small talk with you."

"I wouldn't mind that. I'd love it."

"I'd like to talk with you too. You seem to be a nice person. But under the circumstances, maintaining proper decorum is the right thing to do. It's our job. We should take it serious."

"Guess you're right."

"And besides, I think I like my job. Don't want Harrypaul to give me the boot if he suspects I'm just hanging around your desk and preventing you from doing your work."

"No way. He'll never get rid of you."

"Why is that?"

"Since Sugrim was killed in Micobie, Dennis has become an emotional wreck. He's afraid of his own shadow. He thinks someone is after him."

"Really. Are you serious?"

"Mr. Harrypaul tells me everything. He must think I'm his walking encyclopedia."

Before Wilder had time to respond, a young woman emerged. She was, Wilder thought, as beautiful as ever. Her eyes were mirthful crescents. Medium built, about five-six, with a thin body that appeared inharmonious with her chunky buttocks.

"Frank, this is my friend, Leila," Alma said with an animated smile. "Leila, this is Frank. We work together at the ministry."

"A pleasure meeting you, Leila."

"Same here, Frank. It's such a pleasure."

"Let me get you a drink. What are you having?"

"Brandy with milk," Leila replied.

They talked, listened to music and danced. Something Wilder did not do since the times he hung out at clubs in Newark and Orange. Leila was quite a talker, Wilder observed. She was employed as a manager at Guyana stores and lived with her boyfriend on Charlotte Street in Lacytown.

Hours went by.

Wilder and Alma Rampersaud kept steady heads as they tossed off drink after drink, in small amounts. But Leila Kissoon swished her brandy on her tongue and assumed a judicial expression.

"Frank, I gotta go. Have to take Leila home before she becomes more soused."

"Good idea. Won't be long before she gets there."

It was 1:30 a.m.

The crowd at Buddy's Nite Spot had slowed to a trickle. Wilder stood in the parking lot and watched as Alma and her friend boarded her Morris Minor and headed towards Leila's home in Charlotte Street.

Wilder walked towards the SUV. A man, smashed by cheap whisky, fought to maintain his sense of equilibrium. Despite his inability to stand, he begged Wilder for money with a hoarse alcohol roughed up voice. Wilder tossed him a bill. The man hurled himself in the air and caught the currency with gnarled, banged up hands. A triumphant grin flashed across his face.

Wilder got into the Toyota and headed towards Seawall Road. He stopped at the light and made a left turn on green. He was on his way to Homestretch Avenue into Prashad Nagar. With a bit of luck, considering there wasn't a lot of traffic, he figured he'll be home in about fifteen minutes.

Then he thought of the old neighborhood where he was born and where he grew up until he was twelve. It was Third Street in Alberttown. He made a right on Homestretch Avenue and headed towards Alberttown.

He got there in twenty minutes.

He passed the public school and the playground where he used to hang out with his friends. The building was crumbling into decay. Trees and plants merely existed. But it was also a place where kids from the neighboring community of Queenstown beat his friend with bats, just because he happened to be walking his dog.

Wilder continued along the five square blocks of his old stomping grounds and was surprised to see the nicely painted house where he once lived was then peeling. Old refrigerators and other pieces of unwanted furniture were thrown in a stagnant, unmaintained gully here and there.

Then he thought of Prashad Nagar, the place where he resided. He realized that the shifting demographics of Georgetown were all about the almighty dollar. A different epoch in time.

Wilder knew that the bottom line for him was that there was nothing left of himself in the streets of Alberttown or the house where he went from an embryotic state to homo sapiens. The denizens he passed on the streets during his childhood were not familiar anymore. A few friends who had constantly been in trouble with the police might have long since been dead or vegetating in a prison.

Alberttown to him was long gone.

Except in his mind, where it would endure as long as he could remember it.

Before long Wilder was driving towards his place in Prashad Nagar.

Then he thought that hours sooner, he had run into good fortune by a chance meeting with Alma Rampersaud, away from the Ministry and Dennis Harrypaul.

Something of value, he thought, did emerge from that serendipitous meeting at Buddy's Nite Spot. He remembered her words when she said Harrypaul told her everything. That meant a lot, Wilder figured. The woman might have known a great deal of what transpired after the disaster at Omai. Just by being Harrypaul's executive secretary, Wilder was sure she was privy to a lot of classified information.

Then a feeling of achievement ran through his mind. He felt he was closer to busting the case wide open more than he had ever thought. First he had to work along with Alma Rampersaud. He didn't care if he had to fall in love all over again. Like he was with Sofia McKean in New Jersey. Things would fall apart any damn way he reckoned. Beside he knew that his heart was too hard for love.

Alma wanted to see him again. He wanted to see her too. Tomorrow they were meeting, he remembered. Maybe for dinner or a drink somewhere in the country.

The Land Cruiser responded to his gentle pressure on the gas. With a corresponding surge in power. He made a right on Savannah Road and pulled into a driveway, about the middle of the block.

As Wilder opened his front door, his cell phone rang. He fished the instrument from his pocket and opened it.

Alma Rampersaud.

The name appeared on the screen as it kept ringing. Then he remembered that they had exchanged numbers at the club.

"Hello," Wilder answered.

"Hi, Frank. This is Alma. Did you reach home safely? I was thinking about you."

"Yes. I'm ok. Thanks for the thought."

"Thank you for last evening at Buddy's. I had a wonderful time. Leila said she enjoyed herself too."

"I'm glad. I guess we all did have a good time."

"You're a great dancer."

"You too, Alma."

"Am I really?"

"Yes, you are."

"You're not putting me on. Are you?"

"No. I'm for real."

"Well, it's already Saturday."

"You're sure right."

"And tonight we have a date."

"Yes. I'll pick you up at eight."

"Ok. I'll see you later today, Frank."

"Ok, bye."

When the conversation ended, Wilder walked towards the southern window of his apartment. He opened it and gazed into the night. Moths, beetles and a host of nocturnal insects fought for space under a street lamp. But beyond that, the view of the Atlantic Ocean was spectacular. He had never lived so close to the ocean before and was astounded to see the huge body of water, like a gigantic bowl about to overflow. The slender dark ribbon for a horizon stretched out like the wet black asphalt of the great Pan American Highway.

Wilder breathed in the cool night air. To him it was more intoxicating than the couple of whiskies and ginger he downed at Buddy's.

He secured the window, put his 9 mm under his pillow and crawled into bed.

Chapter 19

It was 6:30 p.m. on Saturday. Alma Rampersaud emerged from her shower, walked into her bedroom and gazed at herself in the mirror. She felt that she had not looked very good last evening. She was thirty-four. But some people at the office thought she looked younger. Despite that, she felt obsessed with the subtle wrinkles that had suddenly made their way by her eyes ever since her last birthday. Then she noticed a small furuncle appearing on her skin. Just above her belly button. She attempted to squeeze it and bring it to a head, but nothing happened. Alma studied herself with a critical eye and thought about men. She wondered why some found her so attractive. She sat on her bedroom stool and applied the body oil to her skin. She laid it on with ease. Its voluptuous fragrance cascaded along the hills and contours of her anatomy like a gentle breeze, yet the rich odor of her hydrating lotion and body wash hung stubbornly in the air.

Then she got up from the stool, again bringing her one hundred and twenty pound frame in full view of the mirror. She felt somewhat pleased with her body. A body that never bore children, although she thought, she needed to do some exercises.

Suddenly a feeling of ambivalence ran over her as she thought of a little town named Helena in Mahaica on the East Coast of Demerara. There was where her story began, she remembered. She was only five. And until that day, fuzzy images of what she was able to recollect, still invaded her consciousness. Like the blood on the pillow where her mother had slept beside her. The cutlass in the hand of her father Mohabir as he walked back and forth in their little house. The words he yelled to her firmly, "Get back to your bed!"

She obeyed. Snuggling beside her mother who, years later, she would understand, was already dead from neck and chest wounds and blunt force trauma to the head. She didn't understand why her father left her alive that September night in 1968, when he wielded his cutlass in a drunken rage and sliced his wife without mercy.

A week later Mohabir Rampersaud told police that he and his wife, Tethrie, had problems with their marriage. But he didn't have any problems with his daughter, Alma.

A year later, Mohabir Rampersaud, forty-one, was executed in Georgetown for the murder of his wife Tethrie. Guyana administered the death penalty by hanging, fulfilling a jury's call for the death sentence, delayed by six months of legal challenges and pleas that Rampersaud's life be spared.

In the coming years, Alma Rampersaud had attempted to make sense from that horrific night in Helena. She fought with her thoughts.

For a long time her story didn't seem real to her. As a girl, she heard muffled conversations at school and other gatherings about the woman who was chopped to death at her home in Helena. But Alma knew she was part of a bigger picture. But only in an uncertain way. It was to her like hearing a precept in someone else's history.

Her uncle and his wife, with whom she grew up in Paradise Village in the East Coast, never avoided the subject, but they also never brought it up, Alma would tell others.

As a pre-teen and teenager, she felt unquiet or troubled around men, especially the fathers of her friends. They seemed volatile and aggressive. Alma used to flinch when people got upset or pushy. She avoided confrontation. She grew up seeing herself as weak.

Once at Paradise Regional High School in her senior year, a teacher assigned the students to write short stories about their lives. Alma did not know if she should include an account of her survival that day in Helena. But eventually she did.

And after nearly two decades, she sat in her flat in Lamaha Gardens and recounted that conversation with her teacher, about the murder of her mother.

After graduating high school she moved to Georgetown where she lived with her mother's only sister in Wortmanville.

She started at the University of Guyana in Ogle that September after high school. Four years later, she earned her BA in business administration. She started as a confidential secretary to Judge Herman Sullivan at the nations supreme court in Georgetown. There lawyers and other judges wanted to go out with her. Other men even wanted to get married and raise children. She declined the offers of most men, but eventually dated two or three during those years. But she swore never to get married or bear any children on account of what had befallen her mother.

After three years in the civil service, Alma made an application to the Ministry of Health for the position of executive secretary to Dr. Dennis Harrypaul, the new Minister of Health. To her surprise, she got the job.

Then came the cyanide spill at Omai Gold Mine in 1995 and about a year later the sudden, unprovoked attack on Sugrim Bhoopat, Harrypaul's aide.

For Alma, the death of Bhoopat at Micobie Village in the Potaro and the manner in which it was perpetrated was only a reminder of her own experience.

With regards to her father, she always knew that whatever sentence he was given was rightly carried out.

And so it was with the Amerindian who killed Bhoopat, she figured. The penalty ran parallel with the crime.

Despite her philosophical approach to all those events, she still battled her own demons.

Day after day.

Chapter 20

Saturday evening in Georgetown, Frank Wilder left Prashad Nagar about 7:45, on his way to Alma's flat in Lamaha Gardens. She lived in a small rented, white-shingled bungalow, anchored between slices of other buildings on Duncan Street.

Wilder parked his vehicle in the driveway next to Alma's car and climbed the stairs. He knocked on the door. Before long it was opened by Alma.

Wilder shot her a quick gaze. Up and down. She wore a slight beige cotton dress with short sleeves. It was split about eighteen inches up the sides. But cut full with a drawstring waist. There was no touch of scarlet in her hair. It was dark and luxuriant.

Her chiseled face, the full breast under the beige cotton dress and her shapely legs, sent a hot wave of desire throughout his body.

"Hello, Alma," he said, looking at her.

"Hi, Frank. You're right on time," was her answer.

Then, he thought he saw an adoring stare in her ebony eyes. The irises striped with subtle, horizontal lines of dark brown and her face moist with appeal.

Wilder walked to the middle of the floor. She was close to him as she stretched up and gave him a kiss on his cheek. He responded by kissing her also on her cheek. Then he felt her breast against him. The front door was still opened. Wilder pushed it shut.

"I shouldn't have done that," she grinned.

"We're friends. Aren't we?"

"Yes. You're right."

"You've got a beautiful place."

"Follow me. I'll show you around."

Then he followed her to the kitchen.

She rented the bungalow from a Portuguese man two years ago. He had it restored. Even the cabinets were refinished. Old appliances were replaced. And the floor was redone.

"It's like a house in a magazine," Wilder said.

Half hour later, they arrived at an upscale restaurant at Regent and Hinck Street. Wilder ordered a rum and ginger for himself while Alma got herself a vodka martini.

"I can't remember when I've been so happy to see someone," Alma said, sipping her drink.

"I'm glad its me you're happy to see," Wilder replied.

They ordered their food.

The waitress said they had a fifteen minute wait. That didn't bother Wilder. He figured he had all the time in the world.

"You don't mind the wait, do you?" Wilder asked.

"Not in the least."

Then she slipped off her shoes under the table cover and attempted to play with his feet with her toes. She felt his ankle holster with the 9 mm Browning. She blinked with surprise.

"Are you supposed to wear your gun all the time?"

"Yes."

"What if it falls out. You can lose it."

"Its strapped in a holster."

"You have a badge too, I guess."

"That's right. Can't be a policeman without a badge."

"Can I see it? Just want to see your picture."

Wilder produced a small leather wallet and pushed it across the table.

"It's gold," she exclaimed, picking it up.

"No. It's brass, covered with gold."

Then she looked at his picture behind the logo of the Police Academy in Eve Leary.

"You're nice in this picture. Do all constables have badges like yours?"

"No. Not at all."

Just then the waitress brought their dinner. Wilder scooped up his badge and put it in his pocket. Then they ate hungrily.

"Why some ranks don't have a badge like yours?"

"Because I'm a special rank."

"How special?"

"I was hand picked to work for the Ministry of Health."

"You mean to keep Dennis out of harm's way when he's in public?"

"Something like that. And speaking of Dennis, you told me last night at Buddy's that Mr. Harrypaul tells you everything. Is that right?"

"Sure, he talks to me freely."

"So was he the one who authorized the firings of those three ranks after Bhoopat was killed? It doesn't mean anything to me. But anyhow, Dennis must have a lot of pull to wield such power."

"Sure, he called Ramesh Doolay, the Minister of Home Affairs, and had them fired."

"Just like that."

"Sure."

"Do you think that President Jagnarine was aware of all that stuff?"

"More than aware. At times, Dennis and the President meet at Le Meridien restaurant and drink rum together."

"Is that right?"

"I swear. I'm not talking with a double tongue now."

"They must be good buddies."

"Something like that."

"What about the ranks who were fired?"

"What about them?"

"Did they get other jobs?"

"No one knows. The government refused to release their names to the news media. Even their addresses are kept a secret. But I'm sure that information is somewhere at police headquarters in Brickdam."

"Really?"

"Sure. I sent the e-mail to Police Headquarters as per Dennis's orders. Their names, addresses, age, date of hire and fire were put in their files."

Wilder wanted to blink with surprise at those revelations, but he controlled his emotions by efforts that sent a spasm of adrenalin rushing through his body.

"Want another martini?" he asked.

"Just one more. Then maybe I'll go home, 'cause I'm already mellow."

Wilder gestured to the waitress. He ordered two drinks. One for Alma and the other for himself.

It was 12:30 a.m.

Alma sat across the table from Frank Wilder. She giggled. She picked up her drink, took a gulp and lay it back on the table.

"I know what you're thinking," Alma said as she looked across her table with dark provocative eyes.

"How can you be sure?"

"You've got a suggestive smile."

He did not answer. But they exchanged looks.

Then he noted a sudden hunger in her eyes and a candid gaze on her face that caused her entire body to blossom into a new fullness.

"Let me take you home," he finally said.

"Sure."

He felt alert and in control as she wrapped her hand around his waist as they headed across Regent Street towards the Toyota SUV.

When they got to Lamaha Gardens, it began to rain. The huge hail-like drops bashed the streets in a violent display of nature's wrath. They got inside her house. She turned on her stereo, popped a CD and drew the heavy curtains across the windows. Wilder sat on a sofa. She turned off all the lights except the one on her coffee table. Then she sat next to him on the sofa and kissed him, full on the lips. He sensed the odor of her breath. Vodka martini. It pleased him. She continued to force the action, pressing her one hundred and twenty pounds of flesh on his big frame. He felt the subtle gyratory movement of her tongue against his. His body reeled with unfettered joy as he responded like a savage beast, kissing her, cuddling her, touching her breasts and yanking her closer to him.

His fingers ran down her dress. Midway. They slipped the drawstring. And with some expended effort he removed the garment. In haste, he tossed it across the polished floor. It lay unnoticed in a crumbled heap of cotton. Then he removed her bra.

After those brief moments, he found his voice. It sounded weak, quiet and constricted. To his own ears.

"Let's go to bed," he whispered.

They got up from the sofa, still hugging and kissing. He got the faint scent of her perfume, her body musk and her hair. Her delicate fingers hurtled up his shirt, undoing the buttons from top to bottom.

He felt a sudden, strange sensation of heat about his lions, as her fingers descended to unzip his jeans.

Wilder looked at her, with salacious eyes. Feasting on the lovely obliqueness of her breasts. Tawny colored organs, ripe with passion with nipples hardened by the call of her emotion.

He picked her up and walked towards her bedroom.

Her bedroom was a combination of young girl and grown woman. There were framed still pictures of Madonna, Michael Jackson and Aretha Franklin on her dresser and vanity. Other pictures of herself in leotards and gym costumes while in high school littered the wall. There was a small boom box on the floor close to her bed. Some toys, fuzzy animals and dolls from yesteryear, lay scattered about her pillow.

She smiled dreamily as Wilder laid her down. In a crouch, he scooped off his underwear. When he looked up, she had also taken off hers.

Wilder ran his hand along her breasts, abdomen and thighs, again and again. Then with firm insistence. It made her utter low mirthful whispers under the influence of her bliss. Again their lips met. Then he lowered himself upon her. Her legs came up surrounding his waist as he started to enter her. Then her lascivious eyes stared adoringly at him before they closed in rapture.

Then his organ, hot and throbbing, as if possessing a heartbeat of its own, ascended her vitals. She moaned softly in her felicity as it opened her up and wiggled deeper and yet deeper into her.

The minutes crept by with reluctance. As if lingering with her delight. Ten. Fifteen. Twenty. Thirty. Forty five minutes.

She whispered his name softly as she sucked in her breath.

Somewhere in the house a clock was ticking. And a giant rumble of thunder on the outside rattled the skies. Random flashes of lightning brightened the land with its violent discharge of atmospheric electricity.

The furious thunderstorm that had begun earlier continued to rage.

And in bed, two distinct human beings were then one.

She ran her hands along his torso and felt his savage energy. Inside her, a heart was thumping against her chest. Her blood was boiling. Then she attempted to place her cupped hand over her mouth to keep herself from screaming. "Oh, Frank. Please, don't ever stop!" Then she was bombarded with orgasms after orgasms of delicious agonies that rocked her body. He breathed words of endearment in her ear. She closed her eyes as if dissolved by his touch. Then she wondered if she was the same genteel, Christian lady she was looking at in the mirror.

Chapter 21

Morning came to Lamaha Gardens.

Alma Rampersaud turned in bed and opened her eyes. Wilder, too, was awake. Naked, he looked even more feral than before. Shoulder, chest, belly and legs covered with hair. He pulled her closer to him, but after a few moments she pulled away. She looked at him as if she was no longer quite sure of herself. Then tears welled up in her eyes.

"Alma!"

"What?"

"You're crying, baby. What's going on?"

"It's my parents, Frank. Sometimes when I remember, it makes me cry."

"What about your parents?"

"My dad killed my mother with a cutlass when I was five. He was a drunk, Frank. A no good son-of-a-bitch. He died on the gallows."

"Holy shit. You never told me this!"

"Not much to tell."

"So both your parents are gone."

"Yes. Gone forever."

He felt unable to talk when he thought of his own situation with his parents.

"It's not fair," she said after a pause.

Then his eyes turned to meet hers. His voice was soft. "Nothing ever is, Alma."

"Why did you say that?"

"My parents, like I told you at Buddy's, are both dead."

"I remember you told me that, Frank. I feel so bad. Did they die naturally or in an accident? Whatever happened?"

"With my mother, it's a long story. I was sixteen years old then, living in Newark, New Jersey. My mother went to her job one day in Brooklyn in 1974 and was never seen since then."

"Why? Did she just run off?"

"No one knows. But the police suspected foul play after they found her empty wallet on the subway line months after she disappeared."

"Was anything done?"

"After fifteen years of investigation, the case ran cold. My mother went missing and was presumed dead, so said the district attorney."

"And your old man. How did he die?"

"He was so heartbroken that he started drinking. He died an alcoholic."

"Oh, I feel so sorry, Frank."

"I'm sorry about your situation, too." There was genuine sympathy in his voice.

Hearing those words from him, she felt a new fondness. A feeling that caused her not to feel so totally alone.

"Whatever happens, Dennis must not know about us," she said after a sudden silence, her face bearing a troubled expression.

"Why is that your concern?"

"This is Lamaha Gardens and people will talk."

He looked her in the eyes. "You're right. I don't want to cause any problems for you."

They continued to meet each other's gaze.

"I'm not thinking about me, Frank. Remember you're a black man, and I'm an East Indian woman. They would not like that. And that's the way they think. I mean the Ministry."

A smile crossed Wilder's face. "That's good to know. Didn't realize they'll think that way about me," Wilder said naively.

"You've been gone from Guyana for a long time. You're almost a stranger in your own country now. Things have changed."

"Guess there's truth to that."

"Dennis is a filthy old man. He makes me feel like dirt, the way he tells nasty jokes in the office sometimes. But that will not stop me from being your woman."

"That's Politics 101 for you. You're in a business controlled by a bunch of filthy-ass old men and faggots. Just gotta go with the flow."

"I know. But he has nothing on me and he knows it."

"What do you mean?"

"Frank, I got more dirt on Dennis than you can shake a stick at."

"How so?"

"Words are going around Georgetown that he had a hand in Sugrim's death."

"You're bullshitting me," Wilder winced as if caught by surprise.

"I won't lie."

"Why would Dennis be involved?" Wilder asked even though he suspected something fishy from the beginning.

"It's a long story, Frank. Look here, I gotta have some coffee. Do you want some?"

"Yes."

She arched her naked body to a sitting position and retrieved her panties that were lying on the floor. Then her bra. Her black disheveled hair was spilling around her face. Making her look promiscuous.

She looked at Wilder dreamily. As if still in a state of trance from her paroxysms after paroxysms of unforgettable orgasms.

She slid off her bed and walked to the bathroom. Wilder couldn't help looking at her. He figured it wasn't just something one could just turn one's eyes away from. She looked so buxom, so lascivious, that her body looked like a pattern of creation that could blow one's mind. When she emerged from the bathroom, she deliberately flaunted her attributes, knowing he was watching her.

"Let me get you the coffee," she said.

Wilder pulled on his drawers and followed her in the kitchen.

Within minutes the coffee was ready. The rich smell of cinnamon permeated the air. They sat at the kitchen table. Then he spoke.

"I'll drink mine with a little milk and sugar. Chocolate brown."

"That's the way you like your women?"

"Skin color means nothing to me."

"Sounds like you know what you're talkin' about."

"You may say so. I've seen them in all color and creed back in the United States. And I learned that what defines a woman or a man is character not skin color."

She got from her chair, leaned over and kissed him. Then she whispered, "Do I have what it takes, Frank? I mean last night. What do you think?"

"You were dynamite."

She giggled. Then held him with a vivacious stare.

Wilder wanted to know the reason for Dennis's involvement with Sugrim's death. And Alma knew what went down. He sensed her reluctance to tell him everything in a straight way. He figured he had to employ patience and tact in his approach to her. The woman should not have the least iota of suspicion that he was fishing for information.

"You were talking about our boss, Dennis," Wilder said. "How was he involved with Sugrim's death?"

Alma swallowed some of her coffee and then cleared her throat. She looked at him candidly.

"Frank, I love you and I trust you. What I'm about to tell you must stay right here. Nobody must know. Else they'll find me floating in the Lamaha canal."

"Sure, you can trust me."

"A week after the cyanide spill, people living around the area as far as Burro Burro River reported that people were dying after using river water. Especially the Amerindians who farmed and hunted close to the rivers."

"And what did the government do?"

"The police stationed in the area reported it to the Ministry of Health."

"And they looked the other way, right."

"Worse than that. They heard that the Amerindian leaders were requesting the US EPA come to Guyana to investigate the spill as a violation and abuse of human rights."

"What they did? I mean the Health Ministry."

"They got government goons to go into the interior and bury the Amerindians who died from the poisoning in mass graves."

"You must be joking. How did Dennis come into all this?"

"He's the Minister of Health and remember, Sugrim was his subordinate. They worked together to pay people to stay quiet and also to conceal evidence."

"The president is aware of all this."

"Not in the least. Every ministry minds its own business."

"Then why Sugrim died?"

"There was rumor that a human rights organization was coming to Guyana to set up a commission to question Sugrim about behind the scenes activities by the Ministry of Health after Omai."

"But then Dennis's name was not even mentioned," Wilder added.

"He got scared," Alma answered. "He knew that sooner or later they would point a finger at him."

"What did he do to get out of that?"

"The man's crafty. At least he thinks he is. He seized the opportunity when the leptospirosis broke out in Micobie and sent Sugrim to hold a town hall meeting to address the issue. Unknown to Sugrim, the three ranks who were his escorts were given orders by Dennis to recruit the Amerindian to kill his deputy. The Amerindian did what he was ordered and paid to do. But he was a liability to the Minister of Health. He knew too much. So the ranks had their orders to execute him on the spot. Right now, there's no one out there who will come forward and say that Dennis was involved in the murder of his deputy."

"Yes, there is," Wilder said.

"Who?"

"The cops who got canned."

"I don't think so. They're running scared and will never come forward and tell the truth. Besides they can be in big trouble themselves," Alma added.

"That's right. They can be held with Dennis for the murders of Sugrim and the Amerindian."

"But who's gonna prosecute them? No one, I guess. This isn't the United States, you know."

"That's for sure," Wilder answered.

"So we're back to square one."

"Yes. But I'm still puzzled as to how you know all this stuff."

"Easy. One day Dennis and Sugrim were in the office. Dennis didn't know he left his desk com opened. I heard it all. They discussed Omai and how people from the WHO and US EPA were coming to Guyana to investigate the disappearance of a lot of people in the interior of the country. Dennis told Sugrim not to say anything that would incriminate them because heads will roll."

"Was Dennis aware that you overheard their conversation?"

"No. Not in the least."

Chapter 22

Weeks went by.

They turned into months.

Frank Wilder and Alma Rampersaud continued their love affair away from the Ministry of Health. For him, the frustrations were beginning to stack up. He didn't need a seeing eye dog to let him know that the woman was genuine. And that she was passionately in love with him. But was it not right for him to return her love? He walked around his apartment, seething, thinking of Alma, wondering how he'd ever been dumb enough to think he was in love with her. Worried more about his mission and the effect a love affair would have on it. Half his mind was on her. The other half on what was going down inside the Ministry of Health.

Wilder attempted to envision the reality of their relationship. That he was a black man and she an Indo-Guyanese woman, was the least of his concern. The woman had a talent, he thought, that subtle erotic power to turn him on. A power so strong, so dominating and overwhelming that, when she lay hands on him in a voluptuous way, all he thought of was sex. He rebuked himself for being unable to keep her distant, out in the universe of his mind. Night after night, she had been offering that stunning body to him. And by heavens, he had been unable to say no. He wondered how in hell life would be if he were to call off the affair.

And what on earth would ever happen when he becomes desperate for it.

That possibility bothered him. And at that moment, he thought he would give anything to be in bed with Alma. Just to be next to her in a room and their clothes off.

Suddenly he wanted her more than anything in that whole wide world. More than money or success in his choice of career. But again, he was investigating her boss, Dennis Harrypaul, for crimes committed after the cyanide spill at Omai. And unknown to her, based on what she had told him about Bhoopat's murder, she might be subpoenaed to testify against the Minister of Health in an international court.

Wilder lit a cigarette, then paced the floor of his apartment. His face drawn in deep thought. What weighed against the story was Alma herself, Wilder reckoned. He knew her character. She drank socially. No drugs. She was stable given her tragic history. And sexually attractive to men as he well knew. But yet to balance the scale in his search for the truth, he had to disappoint her and break her heart.

Still, dammit, Wilder wished Alma was mean. And a bitch and a slut. Then he didn't have to generate one shred of emotion in carrying out his duty.

That would have been a heck of a lot easier, Wilder thought. But unfortunately that was not the case.

He knew that the wall of so-called invincible power Dennis Harrypaul had built around himself and the

Ministry of Health was made of dominoes and was about to come crashing down.

As for Alma, Wilder again wondered about her. Her winning charm, her unchangeable geniality and her warm embrace never disappeared from his mind's eye. How many times had he lain naked besides that beautiful East Indian woman, who had been naked also, enrapt in a tempestuous episode of love making and wished that it would never end?

But grudgingly, he was aware that the chips would fall wherever they had to fall.

He looked at his wristwatch. It was five minutes after seven that Wednesday evening. That night he would be taking Alma to The El Dorado, a club and restaurant situated on the first floor of Le Meridien Pegasus Hotel by the ocean.

An hour later they arrived at their destination. A waiter showed them a table not far from the bar.

The place was busy. The hum of voices from the customers depicted their satisfaction of the food and the surroundings.

"It's nice and quiet here," Alma said. "This place goes crazy at lunchtime. Everybody's in a big rush. But I guess here we can talk."

Wilder wondered what they would be talking about. Momentarily, he felt a jab of concern in his stomach. Was she aware of something? He kept his composure.

They ordered drinks and trifled over the menu. Alma's mood was upbeat as always.

She stared at him for a moment.

"What are we going to do, Frank?"

"What do you mean?"

"You and I. What will ever happen?"

"You ask tough questions."

He stood up, grabbed her hand and led her to the dance floor. She went into his arms as if under a spell. At first her legs were stiff as she moved contrary to the music. He pulled her closer to him and whispered in her ear, "Try to relax. You're allowed to have fun like anyone else."

She giggled and they settled into the rhythm of the music. Then she rested her head on his shoulder. "I want to thank you."

"For what?"

"You've made me feel like a woman."

"I don't understand."

"Today a man called and asked to speak to Dennis. I asked him for his name. He refused, only to say that he was a former cop and he wanted to talk to Dennis urgently. I put the call through to Dennis. Minutes later Dennis called me in his office and almost went on a rampage because I put the call through to him."

"What!"

"Yes. He yelled profanity at me."

"But the man wanted to speak to him. You did the right thing."

"Frank, I was so hurt. I felt like dirt."

"Don't take it so hard."

"What that man is putting me through after Sugrim's murder makes me want to quit."

"I know you do. So the caller said he was a cop."

"Yes."

"I won't be surprised if it's one of the guys he fired."

"That's possible."

"But why does he want to talk to Dennis? Something isn't right here."

"You're the cop," Alma answered.

Wilder did not respond. He eyed the waiter bringing their food. Curried mutton and rice and Banks beer to wash it down.

They enjoyed their dinner and hours later, Wilder took the woman to her home. She held his hand when they got out of the vehicle.

He followed her into her house. Before long she was unzipping his trousers. Then things seemed to occur automatically. They were on the sofa, his jeans hovering by his ankles and her legs around him. She was without underwear. They both got intertwined. Each wishing those minutes would last forever. Then they experienced their orgasms with her sobs and cries mixing with his whispers that reached the very zenith of ecstasy. Then they lay spent in each others arms, breathing heavily.

A few minutes after one am, Wilder kissed Alma and left for his home in Prashad Nagar.

Then he thought of the verbal outburst Dennis Harrypaul had with her. Wilder figured there was some hidden reason as to why the man blew his top. But why did an ex-cop try to contact Dennis? Where was the connection? Wilder viewed the whole scenario with a lot of skepticism. Not of Alma, but of the hidden agenda that the Minister of Health seemed to nourish.

He thought of the ranks who were fired by Dennis after Micobie. Might it be that the constables knew something and one of them was trying to make contact with the Minister of Health.

Wilder tried to look at the bigger picture of what might be going on and he felt some concern for Alma. A gracious and elegant woman struggling to put aside a tragic past. But unknowingly caught up in a tangled web of political intrigue where Wilder figured unprincipled, fraudulent and underhanded activities were the order of the day. As a guard at the ministry, it didn't take him long to know that he was part of an environment where two-faced, sneaky and self-serving business cronies with opened hands, wandered in and out of the Ministry of Health, during working hours, bitching about the sorry state of governmental affairs and moaning about their phenomenal efforts to stay ahead of a losing economic game.

His hands tightened around the steering wheel of the Toyota SUV as he felt a gnawing sense of powerlessness. Before it laid a strangled hold on him, a half-hearted smile crossed his face. Then an idea ran through his mind. He had to go back to the Ministry of Health, unknown to anyone, to find information on the ranks involved in the Micobie incident. He made a U-turn at the corner of

Lamaha Street and Vlissengen Road and headed north on Vlissengen Road.

The heavy stench of the zoological garden littered the air when he approached Robb Street. He turned on Brickdam Avenue.

It was 1:30 a.m. when he parked the vehicle at a night club on Hadfield Street, about half a mile from the Auto Supplies store. He figured that to police cruisers his vehicle would look like it belonged to a club patron. He decided to walk about three quarters of a mile to Brickdam Drive, where the ministry was located. Walking didn't bother Wilder. He knew it was good for his heart and his gut. But then he reasoned that walking that night was out of the question. He flagged a minibus. It dropped him where he indicated. A block from the building where he was headed.

Wilder walked down the street, his 9 mm tucked in his shoulder holster, acting like a man going about his business.

The Ministry of Health was at the corner of Parker Street and Brickdam Avenue. Across from the ministry there were a couple of commercial buildings without night watchmen. The ministry occupied the corner and four other empty lots for employees' parking. Immediately after the parking lot there was a stationery store that was closed for renovations. Wilder did not worry about anyone observing him.

Quickly he pulled a dark hood from his pocket and slid it over his head. He made certain the eye holes were adjusted and he could have seen well. Then he approached the walkway where a lone policeman was on duty in a sentry box. Wilder prowled closer to the guard and found the man asleep. A misshapen, short, fat and bald headed creature, Wilder thought, as he watched the constable locked in slumber, snoring like an ant-bear. Wilder tapped the rank gently on his shoulder. The guard responded with an astonished cry. But before he could have opened his eyes, Wilder knocked him cold with a blow to his right temple. The man fell heavily across his desk.

Quickly Wilder donned a pair of latex gloves he fished from his pocket and scampered up the stairs. He entered the six digit security code on the key pad close to the door knob and neutralized the alarm. With a pen flashlight in hand, he entered Alma's office. His eyes raking the surroundings. With a catlike grace, he tottered through a passageway that led to Harrypaul's office. The door was locked. He pondered for a few moments as he realized he did not know the code. He backed up and made his way again to Alma's office. He looked at the computer. A windows XP desktop. He had some hands-on computer laboratory training during his early days with Interpol. He picked a category and clicked twice to access the recycle bin. Quickly he downloaded the information on the Micobie incident. The names and addresses of the three constables fired by the Minister of Health were before his eyes. He grabbed a scratch pad and jotted down the information. Then he logged off and scrambled down the stairs.

The guard in the sentry box was beginning to regain consciousness. In a dreamy way, he was half aware of what was going on when he felt the perception of a hooded figure before him. His bullfrog face tightened in a hard knot as he felt a gnawing sense of helplessness. It was about to strangle him when Wilder hit him again with a parting shot. A vicious haymaker. That time on his left temple. The man groaned as a snarl of agony traversed his face. Then he lumbered back into unconsciousness.

Wilder ripped off his hood and gloves. He threw them into a trash can that was standing on the sidewalk. Then he left the scene on nimble feet.

About four blocks from the building, he got into the SUV and drove slowly towards Prashad Nagar.

Chapter 23

Uitvlugt was a Delta town nestled close to the mouth of the Essequibo River on the wide Atlantic Coast. It was set amidst canals, forest land and rice paddy fields that stretched between Vreed-en-hoop and Parika, forty-five miles west of Georgetown.

Since the closing of the sugar cane factories decades ago, peasant farming was not king anymore. And the livelihood of Uitvlugt dried up. Farmers then became fishermen. But the Omai spill drove them into extinction when demand for seafood declined. People were in awe. The town became an empty shell, weed covered and fettered under the heat of a relentless sun. The conditions betrayed no hint of an economic meltdown that locals thought were exacerbated by what they referred to as the "Omai Overspill."

Then one evening in late 1995, a carpetbagger from Georgetown boarded the Vreed-en-hoop ferry and crossed the Demerara River. He called himself an investment broker and set up shop on Marabunta Road, on the fringes of Uitvlugt. He beseeched the populace to seize the moment in time and make their God-given land work for them.

And despite the illicit nature of the venture, most people took the man's advice.

They manufactured and distributed bush-rum. An illegal whiskey.

Then Uitvlugt became prosperous. And an artifact. No longer a down-on-its-luck community. Or a threadbare town whose only reason for existing was as distant a memory as the region's agricultural roots. It stood out like a jewel.

And everyone knew that Uitvlugt's upward mobility was centered on a shady base. The mayor, town council and the police were all paid to look the other way.

One evening, Frank Wilder came to Uitvlugt. He was looking for a man named Oscar Bascom, a former constable of the Guyana Police Force. Bascom's name had showed up on a list of three ranks who were dismissed from the force after the infamous attack on Sugrim Bhoopat that left him dead at Micobie.

Wilder intended to question the man or the others who he felt had some knowledge that would expose government officials for illegal activity.

He pulled the SUV to a stop on Hibiscus Avenue. It was four p.m. on Saturday. He walked across a bridge that spanned a small canal and entered a candy store. He ordered a Pepsi and a pack of cigarettes.

The man behind the counter didn't look Wilder in the eye. He got what was ordered and shoved it across the counter.

Wilder paid the man.

"Where is McDaniel Street? Have any idea?" Wilder asked.

"Yes," the man said, pointing to his right. "Continue on Hibiscus Avenue for about two miles. McDaniel Street comes up on your right."

"Thank you."

"Don't mention it."

Wilder continued west on Hibiscus Avenue and followed the man's direction. Traffic was heavy. Within twenty minutes he turned on McDaniel Street. He was looking for ninety eight. Then the traffic came to a crawl. Wilder was content to watch the diversity of the vehicular traffic. Donkey-carts, mule drawn wagons, mini buses, cars, trucks and bicycles dragging along McDaniel Street as they tied up traffic behind them. As far as Wilder knew, some cars were modern, but most were fifteen- to twenty-years old with homemade mufflers and bumpers. Other vehicles appeared to have been well cared for.

Ninety eight McDaniel Street appeared to Wilder's right. He pulled the vehicle to the curbside and parked it.

The building was a 19th century coral-stoned bungalow sheltered between mangoes, tamarind and star-apple trees. It's gentle backyard elevation allowed one to watch the fishermen go out to the ocean and return with the day's catch.

The southeastern breeze from the Atlantic Ocean wafted continuously throughout the atmosphere. Wilder thought it was fresh and cool.

Wilder moved his body with a resolute carriage towards the door.

He knocked twice.

There was no answer.

He knocked again. That time even louder than before.

Then he heard footsteps.

A small woman, shrunken with age, opened the door. She threw Wilder a funny glance. But he held her with a non-provocative gaze.

"Good afternoon, ma'am. Are you Mrs. Bascom?"

"Yes, I am. What can I do to help you?"

"I'm here to see an Oscar Bascom," Wilder said, flashing his badge. "Is he your son?"

"Yes. That's my son. Come on in."

Wilder followed the woman inside and closed the door behind him.

She ushered him to the den.

It was a funny little place, Wilder thought. Possibly done over to her liking. One wall was all mirrored with an old organ against it. Nearby was a record player with a shelf of LPs. The floor was mahogany with a coffee table standing in the middle. Two large sofas adorned with blue slipcovers rested against the other walls.

The remainder of the furniture comprised of a dresser, a cabinet and a small conference table. Old copies of books like *Coral Islands*, *The Three Musketeers* and *Black Beauty* along with a past edition of the Sunday Chronicle were on the table.

"May I sit?" Wilder asked, his voice filled with humility.

"Sure."

"Where is your son, Oscar? Would like to speak to him."

"He's not home now. But is there something wrong?"

"Not quite. Just want to discuss his dismissal from the force after Micobie."

"Oh. That was terrible the way they treated Oscar. He was just doing his job."

"I see your point."

"My son did what he had to do as a police constable."

"Sure. But where can I find him now?"

"He spends some of his time at the Cashew Club. Since he lost his job, he's been depressed. I'm afraid he'll turn to drinking."

"Where's the Cashew Club, Mrs. Bascom?"

"All the way up McDaniel Street, then turn left. It's a green building about a mile down a dirt road on your right."

"Thank you, Mrs. Bascom."

"Well, officer, I sure hope you'll be able to let him know that he still has a life ahead of him. And he should quit worrying about that job he lost."

"I'll try my best to talk to him."

Wilder got into his SUV and headed north on McDaniel Street. The address was in the heart of a thickly wooded area. A left turn off McDaniel Street put him on Patterson Avenue then he made a right onto a dirt road and followed it for half a mile.

A neighborhood bar and grill called the Cashew Club lay sprawled between a grove of coconut palm and jamoon trees. The place seemed an unlikely spot for good food and crooked people. But the Cashew Club had them both. Cars were parked everywhere.

Wilder parked his vehicle and walked into the place. He saw a sign standing by the door. "Anybody representing a law enforcement agency, present yourself." The idea was that if one was a cop and didn't present oneself as a cop, he or she would be somehow guilty of entrapment because there was knowledge that the club sold and distributed bootlegger whiskey. But Wilder thought that was a bunch of bull. But it did help to create some nervousness on his part.

He snaked his way between the crowd and made it to the bar. Great gusts of ribald laughter could be heard everywhere. Men spoke to each other with brutal, wretched and obscene language.

"What's your order, man?" the bartender asked with a nasal whine as he examined Wilder's face owlishly.

"Beer."

"No bush oil tonight," the bartender said, meaning illegal whisky.

"Later on. Got to get warmed up."

"What's your beer? Heineken, Banks or Miller? Take your pick."

"Banks."

"Good man. Buy local. Keep the economy moving."

"That's for sure."

Wilder paid the man with a hundred and sipped on his beer. His eyes scanned the place with a furtive look for anyone fitting the description of Oscar Bascom. An ex-cop well in his forties, about six-feet-two-inches tall with a thin face, dark complexion and a bulbous nose. He sported a crew cut with a neat, thin line for a mustache.

A hooker spotted Wilder's twenties on the bar after the bartender slid his change towards him. She broke away from her sleazy friends and ordered a drink from the bar. She stood close to Wilder, holding him with an angelic smile. Then she ran her tongue slowly over her lips.

"Lonely today, handsome?"

"At times," Wilder answered without making eye contact.

"At this moment?"

"No."

She turned, faced him and gave him an obscene giggle. Then she popped a butt between her rubied lips. Before she found her igniter, Wilder struck his lighter and lit her cigarette.

"A real man," she tittered.

"I'll be more real if you tell me where Oscar is, my long lost buddy."

"You mean Oscar Bascom, that fellow who still thinks he's a constable. He hangs around the Cashew Club expecting people to feel sorry for him."

"Yes, that's the man."

"I'll point him out for twenty."

"You got a deal."

Wilder gave the woman a twenty and pocketed the remainder of the currency. He followed her through the crowd and across the floor. Making it look nice and natural like meeting a hooker and buying her a drink.

They passed a pair of jelly-bellied businessmen who swapped jokes while they staggered about in a semi-drunken state. A small hallway led to a back room, nestled with a greenish-red light from a ceiling fixture. A man sat at a table with a glass of whiskey before him. He fitted the description of Oscar Bascom. A TV on a shelf close to the ceiling was broadcasting the six o'clock news. Nearby, two men and two women pickled by bootlegger whiskey were about to fall from their chairs.

"There is Oscar Bascom," the woman said to Wilder in a flat voice. Then she stormed back towards the bar.

"Hello, Mr. Bascom. How you're doing?"

"Good. And who the hell are you?"

"I'm Frank Wilder, a fellow cop like you used to be."

"No shit, Mr. Wilder. What brought you to this damn rathole?"

"I bring you news, man."

"Wilder, the last time a rank brought me news was when the Minister of Health decided to dismiss me. After fifteen years of meritorious service, that asshole for a Minister of Health fired me."

"Well, I'm here to talk about that. I think Dennis Harrypaul is in big trouble."

"Man, I wouldn't be surprised."

"Since the Omai spill, we all know that Amerindians were demonstrating and complaining that they were not treated fairly by the government and Wadpole Corporation."

"Yes, I know that."

"Some of the leaders called on Interpol, the WHO and the US EPA to come to Guyana to respond to their complaint."

"That was going on for a long time."

"Now apparently Dennis Harrypaul and Sugrim Bhoopat conspired to cover up information relating to the deaths and diseases of many people associated with the cyanide spill in neighboring communities. When they heard that these foreign agencies were coming to Guyana to question them, they panicked. Especially Harrypaul. The man was walking around like a zombie, because he knew Bhoopat could be an easy-to-crack, flip-flop stool pigeon under questioning. So he decided to rub him out. And, Oscar, that's where you and the other two ex-constables come in the picture."

Oscar Bascom looked a hard man. The kind Wilder had seen around for years working in law enforcement. His most passionate hatred was the one he nourished for Dennis Harrypaul. Wilder sensed the indignance in the man's voice and the murderous look in his eyes.

"That bastard Dennis. He gave me five thousand dollars to pay the Amerindian at Micobie to kill Mr. Bhoopat. That I did. But the Amerindian did not know according to plan that the other two ranks and I were supposed to gun him down to make it look as though we were doing our duty. Dennis told us he wanted both Sugrim and the Amerindian dead," Bascom said.

"I guess Dennis wanted to be sure that no one was around to point a finger at him."

"Maybe not. Myself and the other two cops knew what happened."

"How much money he paid you fellas to do what you did?"

"Each of us were to get one million dollars. Dennis came to our homes and told us himself with the understanding that he was going to give us each the money when the air was clear."

"Did you get that money?"

"Never. I called the Ministry sometimes but he never talks to me."

"I know you feel like shit."

"Like hell, I do! Didn't mind not being paid money. But when the man had me and the others fired, that was the icing on the cake. I wanted to go to the Ministry and fuck him up."

"Are you willing to testify before an international court against the Minister of Health on what you know in this matter?"

"Sure. I want to see the fucker on the gallows. Even if I have to do time for my involvement."

"And about your buddies, you think they'll cooperate like you're doing?"

"Damn right, they will. Those men are angrier than I. They want Dennis's balls."

"But there's another matter that has to be ironed out."

"What's that?"

"You have to be aware that you and your buddies were involved in a conspiracy to murder the Amerindian and take bribes."

"I know that, Wilder. I'm ready for anything!"

"Your offense can be downgraded with your willingness to testify against Dennis and others for crimes against humanity."

"I'll agree to that. And there are others."

"Who are the others?"

"Wilder, there's a shit load of people involved here, starting with the cyanide spill, Burt Samuels's murder in Georgetown, the secret disposal of bodies dead from drinking tainted water in Tumatumari, Konawaru, Mahdia and other surrounding communities and ending with some top notch names in government. The administration hired thugs to do its dirty work. The mining company also paid government officials big bucks to run Amerindians off their land by eminent domain. The reason was to extend their business ventures."

"Oscar, that's heavy stuff, man."

"Wilder, I know people who saw it all. I got no reason to lie."

"Who has the Omai files?"

"Dennis, of course. He got the files in his office on Brickdam, locked in a safe. His secretary knows everything, man. That's the word."

"Ok, Oscar, I gotta talk with your buddies. Don't mention this conversation. Say nothing to no one."

Both men stood up and faced each other.

They shook hands.

Wilder's face bore an accomplished expression.

"You can depend on me, Wilder. This conversation stays right here."

"Ok. Keep it that way. And be on the lookout. Someone from Georgetown may be around to talk to you sooner than later."

It was 10:30 p.m. when Wilder left Patterson Avenue and got on McDaniel Street. Uitvlugt had put on her nighttime lingerie of dazzling lights.

He lit a cigarette and sent a puff of smoke into the cool night air as he headed to Prashad Nagar on the outskirts of Georgetown.

Tomorrow, he figured, he had to make a trip to Cornelia Ida and LeBonne Intention, both towns on the East Coast of Demerara to talk to the other two ex-cops.

Then he thought of his dialogue with Bascom. The man knew a lot of what was going on within government circles. Then Wilder remembered Bascom's words when he said that Dennis's secretary knew everything. And he wondered whether or not she knew of the existence of the Omai files locked away in a safe on Brickdam.

He figured then he had to come clean with Alma. He had to let her know what was coming down the pike. For two reasons, Wilder thought. One being he wanted

139

her to help him crack the safe. And the other was that he might be in love with her even though he was reluctant to admit that, even to himself.

Chapter 24

The news of Friday's assault on a rank on duty at the Ministry of Health only infuriated Dennis Harrypaul. The Stabroek News stated that the constable was beaten within an inch of his life, by a tall hooded intruder wearing vinyl gloves. The injured man was later found by a policeman in his cruiser who saw the guardhouse empty and became suspicious. The rank was found unconscious lying on the guardhouse floor. He was rushed to the Public Hospital in Georgetown by emergency medical service. An emergency room doctor said during a news conference that the guard was knocked unconscious by blunt force trauma to his temples. He suffered a serious concussion but was expected to survive.

Later that morning Harrypaul arrived at his office at about 8:40 a.m,, appearing tired and edgy. He looked as if he has had little or no sleep the night before. Exhausted, because in the back of his mind, going over and over again, were the events of early Friday morning. Those were Wilder's thoughts. Harrypaul later told his secretary that what made him livid with rage was the failure of anyone to see who it was that attacked the constable. He told Alma he wished he knew the attacker's motive, because he was in fear for his own life.

The following evening, Wilder lay with his hands behind his head, staring at the ceiling in Alma's place. She was lying besides him, gently snugging her body closer to his as if she wanted to get inside him. Earlier they made love on the sofa, then in the shower and finally the conventional way, in her bed. Then she cooked for him. Dhell and rice with stewed cod fish. And with each passing moment, he was beginning to love her more.

He knew what love was. He has been there before. The emotional wear and tear were too much for him. It made him swear that he would never make the same mistake again. But he wanted to argue that Alma was different. Because of that he figured he had to tell her the truth about his real reason for being in Georgetown and working as a constable. But how on earth was he to begin? It was very difficult for him to lie there besides her and contemplate a way to start a discussion that might just rent them asunder. Forever, maybe.

But that was a chance Wilder was prepared to take.

As he opened the bottle of wine, she looked into his eyes.

"You're quiet tonight, Frank. Is it me?"

"There's something I must tell you."

"Is it serious? Are you leaving me for someone else?" she questioned, her face growing pale.

"No, I'll never do a thing like that."

"Then, please tell me what's on your mind, honey."

"I'm not just the constable you and everybody else think I am."

"What on earth are you talking about?" Her voice was cracking. She held him with a despairing look.

Then he dropped the ball on her.

"I'm an agent of the US National Bureau of Interpol."

"What sort of business is that Frank, and why are you working as a rank for the Guyana Police? Someone here is crazy and I know it isn't me!" A rush of heat blanketed her face as she searched for answers without giving him time to finish his sentence. Her eyes widened with astonishment.

"Alma, I'm in touch with your feelings. Please hear me out."

She rose from her chair and began to sob.

Through her sobs she whispered in an imploring voice, "Frank…please don't lie to me. Don't ever leave me here alone."

Then he explained everything to her, starting from the time he landed in Georgetown and ending the night he knocked the guard unconscious in Brickdam.

She held his hand and led him to her sofa. They both sat. By then she had almost dried away her tears, but her eyes took on a wounded look.

"What did you do after you broke in the office?"

"I downloaded the information on the three ranks who were fired by Dennis."

"Did you go to see them?"

"Yes. All three men said Dennis set up the murder of Bhoopat and the Amerindian at Micobie. He went to their homes and promised to pay them each one million."

"That's right, Frank. I told you that, remember. I heard that conversation on the desk com."

"Right, and the ex-cops said they'll be willing to testify against Dennis and all other people involved in crimes against humanity after the cyanide spill."

"Those cops might be afraid to testify against top notch people because of reprisals."

"They gave me their word. Besides the trial will be held in an International Court of Justice."

"I guess that's the reason for having someone from abroad do the investigation. Someone with an objective view of the whole case."

"That's why I was sent here by Interpol."

"Who requested Interpol?"

"The Guyana Organization for Indigenous Peoples (GOIP). You might have read about it in the newspapers that GOIP contacted the WHO, US EPA and the US National Bureau of Interpol after the cyanide spill. GOIP felt that the government and the Wadpole Corporation bungled the cyanide spill investigation. Things heated up after Burt Samuels, GOIP leader, was shot in Georgetown during a peace rally."

"I read about that incident in the newspaper. Do you know that people are saying that the government was behind that man's death?"

"I wouldn't be surprised. But all will be told at the trial."

"Frank, I'm curious to know why you joined the Police Force here in Guyana."

"When I got to Georgetown, my orders were to infiltrate the mining company, observe the day to day operations at the sites and note any operational impropriety that might have contributed to the disaster."

"So you had to join the mine security force."

"Yes. That was my only chance of seeing what was going on."

"Did you notice anything out of the ordinary when you worked security at Omai?"

"Yes. Cyanide tailing ponds were not properly secured. And the company was not sticking to its agreement with the government of releasing the slush slowly into the Omai River. All that and much more will be in my report to the US EPA."

"You were lucky to end up at the Ministry."

"That's after I captured those bad guys who were stealing reclaimed carbon from the mine. Wadpole's CEO took me in Georgetown to work for him and the rest is history."

"You sure can handle yourself."

"In my job, if you don't have those skills, you'll be dead."

"What will happen to us when all this is done?"

"We have to stay together."

"Frank, I tried hard to convince myself that our meeting at Buddy's was going to be nothing significant. When we saw each other the subsequent evening, I was a fool to think that people fall in love so quickly. I tried hard as hell to dispel you from my thoughts. But I couldn't. I love you from day one."

Then they gazed into each other's eyes. As if awaiting a signal from the other. Each becoming more confident that their affair was not based on a one-sided love. Or a fleeting tempestuous ideal. They were in love.

"I love you too, Alma."

"But before, you answered my question very vaguely when I asked what will become of our relationship?"

"It will continue, despite the odds."

"Meaning?"

"Tomorrow at work, I want you to get a copy of the Omai files from Dennis's safe when he's out on lunch."

"That's easy. I know the combination. Then what happens next?"

"I'll deliver them to my boss, Fred Nelson, in Eve Leary. That's all the tribunal needs to make a case against those who were responsible for human abuse after the cyanide disaster."

"Frank, that wouldn't work. Dennis will not rest till I'm dead."

"No. He wouldn't be able to hurt you."

"What makes you so sure?"

"You won't be here."

"You're so right. I'll be in my grave."

"I don't mean that."

"What do you mean?"

"You will be leaving Guyana with me. We will fly to Barbados then to New Jersey."

He stirred her imagination. Her face brightened as his words seemed to have flavored her senses.

"When, Frank?" she screamed as she shot upright from her seat. "You've got to be kidding me."

"No. I'm not. You've got a few days to handle your affairs. Then we'll be out of here."

"I don't believe what I'm hearing!"

"It's true. You're going to America with me."

Chapter 25

St. James, Barbados.

Three days later a Guyana Airways 747 touched down at Grantley Adams Airport. Amongst the passengers were Frank Wilder and Alma Rampersaud.

They made their way through customs and immigration and rented a Morris Minor from Hertz car rental.

The vehicle rolled along Black Rock Highway in St. James. A neighborhood where class lines were blurry. A place where one was likely to see a Lexus or a Cadillac as a broken down Toyota with its bumper or fender missing. Some houses were elegant and well maintained while others were less gaudy and unadorned.

Wilder pulled the car into the parking lot of the Paradise Villas, a resort at the western tip of the island almost hidden between a small forest of bamboo, breadfruit and mango trees.

Later that day the two drove to Bridgetown, the capital and dined at a famous eatery situated in the Cave Sheppard shopping center.

It was the following Monday in Georgetown. Frank Wilder and Alma Rampersaud did not report for work at the Ministry of Health. And Dennis Harrypaul was red hot with anger.

He instructed his aides to call Alma's residence in Lamaha Gardens. But there was no response.

The phone rang out.

The same thing happened when Wilder's residence was called in Prashad Nagar. No one answered.

Harrypaul dispatched constables to the respective homes as an immediate response to the failure of the two to report for work. The ranks found that their places of abode were empty. The furnishings to Alma's house were missing. Her front door was left ajar. The key lying on the floor.

Meanwhile, at Wilder's residence only his suitcase was missing. All the other items that were there when he got the furnished flat were intact. The keys to his door and the vehicle were on a table. The Toyota was in the driveway.

And back at the Ministry after the ranks notified Harrypaul that, apparently the two had absconded, he ranted and raved, kicking over chairs and tables in a wild display of anger. "Don't ever mention the word apparent. Apparent, my ass, constable! I deal with facts. If you don't know a fact about the whereabouts of Alma, don't say shit!" The Minister's voice boomed out in stentorian bellows. "I don't give a fuck about Constable Wilder. His ass can go back to America. But I'm worried about Alma!"

"What do we do, sir?" asked a nervous employee.

Harrypaul beckoned the man in his office and closed the door. His voice was thick with bitterness. "We have to find her. She knows too much. Call the Central Intelligence Department and have them put an all points bulletin on her ass. Put one out on the constable too. We got to get to the bottom of this damn nonsense. They must be into something big."

"But, sir, there's nothing we can do," the man said with a strangled voice. "They have the law on their side. It is not illegal for a person to quit his job in this country."

"Mr. Bachan, you got more stones than a chigoe nit. I interpret the law here!" Harrypaul shot back fiercely. "I know what she's capable of doing, not you. Obey my order, now!"

A week later a detective visited Cheddi Jagan's Airport. After checking the list of all airlines that left Guyana two weeks ago, he found out that Frank Wilder and Alma Rampersaud had left Georgetown for Bridgetown, Barbados.

Agents of the Central Intelligence Department immediately contacted Grantley Adams Airport. They were told that the couple stayed at the Paradise Villa Resort in St. James for two days and subsequently left on a flight to Newark, New Jersey.

The following morning Dennis Harrypaul was up early. He was nude and the gray light of dawn made him look like a lost penguin in Antarctica. He seemed to be wandering about the room almost aimlessly and without purpose. He paused before the dresser and suddenly there were two of him. One standing inside the mirror and the other out. Harrypaul didn't look at himself. Instead he picked up his underwear, looked at it for a moment, then he put it down. Then he turned and looked at his wife in bed.

At 6:30 a.m. the telephone rang.

He picked it up.

"Who is this?"

"Sorry to bother you this early, sir. This is Detective Kilkenny from CID. We got news about the pair of your employees who disappeared ten days ago."

"Good," Dennis answered. "I hope they have the bastards in handcuffs!" he added in a cold, icy voice.

"No, sir."

"Kilkenny, what on earth do you mean?"

"They're not in Guyana, sir."

"Where on earth are they now? You poor excuse for a detective!"

"In the USA, sir."

"Holy crap!" Dennis responded in an irate voice. "How did you cocksuckers allow that to happen. Heads will roll. You hear me! Heads will roll!" Then he slammed the phone on the receiver.

Chapter 26

In 1997, an International Criminal Court with its seat in Trinidad and Tobago, handed down a criminal indictment against Dennis Harrypaul, Minister of Health and others involved in atrocities after the Omai spill two years earlier. The charges were made following a criminal investigation involving Interpol or the International Criminal Police Organization.

The court exercised its jurisdiction because the government of Guyana was said to have reserved its right to oppose any trial involving any of its ministers. But in reality crown prosecutors might have been reluctant to prosecute anyone in a position of great power and authority.

Like Dennis Harrypaul.

When the trial began, the International Tribunal wasn't just investigating the most senior and highest ranking member of President Charles Jagnarine's cabinet. It was also exposing a bullying, nepotistic and clannish political culture that had flourished across the land for decades.

Despite its size in square miles, Guyana was home to less than one million people. As a consequence, the political ballpark was dominated for years by a small group of players, including Charles Jagnarine, president and commander-in-chief, his wife Jane and various protégés followed by paid political goons. And bringing up the rear were the Ministers of Health, Home Affairs and Lands and Mines.

In the following weeks, the investigation into the handling of the cyanide spill widened when a mining company executive pleaded guilty to bribing government ministers.

Then the Omai files were taken into evidence. Its classified information snared the Ministers of Health, Lands and Mines, Home Affairs and other officials who had until then wheeled and dealed with impunity.

Dennis Harrypaul came under heavy scrutiny for his implication into the death of Sugrim Bhoopat and the disposal of corpses around the area of the spill. One judge sitting with the tribunal accused the Minister of Health as being long regarded as a turbulent and outrageous man.

"You and the man you ordered to be killed were relentless in your quest of conspiracy to obstruct justice!" the jurist boomed as he thundered his gavel on his desk.

Then in a sudden surprise during the proceedings, news came from the legislative council that President Jagnarine wanted to be heard. In return for trade and closer economic ties with the US, he would allow other officials to tell what the government knew and might have misrepresented during the Omai mining incident of August, 1995.

One opposition party leader said that the president had to be pressured into that initiative by the US EPA, the WHO and Amnesty International. But he denied those allegations.

The ex-ranks involved in the Micobie incident were in custody and were later called to testify. So, too, were various persons charged with the disposal of bodies in the interior.

After four months, the International Court had reached a verdict.

Dennis Harrypaul was found guilty of crimes committed against the indigenous peoples of Guyana following the cyanide spill at Omai. He was sentenced to life in a Netherlands prison.

He was also found guilty of murder and conspiracy in the death of his aide, Sugrim Bhoopat. And also conspiracy in the disposing of corpses in the Essequibo. For those offenses, he was sentenced to twenty-five years which were to run concurrently with his life sentence.

The Minister of Lands and Mines retired in disgrace for taking bribes from Wadpole's officials in exchange for mining property grants. Alfred Etwaroo was fined two million dollars and all his property was confiscated.

The Minister of Home Affairs was jailed for fifteen years for conspiracy in the brutal execution-style murder of Burt Samuels, GOIP leader and several other supporters.

The ex-constables were each handed ten years, five of which were to be served in a Georgetown prison and the other five on probation for their cooperation with Interpol agents and the tribunal. But the hired crews that disposed of the bodies in the Essequibo were each given twenty years.

The Wadpole Corporation was fined one and a half billion dollars for compensation to all persons who had lost lives, livelihood or who had suffered irreparable damage, both physical and mental as a result of the cyanide spill. Wadpole was also fined five hundred million dollars to be placed in a fund for cleaning up the effected area. The company had to remain under strict censorship.

That same year President Jagnarine retired from public life after he failed in a fourth bid to become President of the Republic of Guyana. He later died from unknown causes.

Meanwhile in the United States, Frank Wilder and Alma Rampersaud were alive and well, living in Morris County, New Jersey. Two months ago, they were married in Las Vegas, Nevada.

Recently he resigned his position as an Interpol agent. And those days he could be found working as a correction officer at the East Jersey State Prison.

His wife worked as a secretary for a major pharmaceutical company.

They are expecting their first child.

THE END

Author's Biography

Ronald Selvon Seales was born in Little Diamond, Guyana. He came to the United States forty years ago. At present he is a retired Registered Nurse. Mr. Seales holds an Associate Degree in Applied Science in Nursing, a Bachelor of Arts in English Literature and a Masters Degree in Creative Writing. The *Omai Files* is his fifth novel. He is also the author of *The Mango Tree, Blackwater, Cry of the Blackbird,* and *Return of the Arawak.* Mr. Seales lives in an Atlanta suburb with his wife, Claudette.